The RETURN OF ORPHEUS

more personal essays about music and the brain

by Lee R. Harris

Dedicated to all of those who have brought music into my life and to my wife Terri, who is my song.

ISBN# 978-0-9861320-8-7

Keywords: 1. Music 2. Brain 3. Psychology

page

CHAPTERS

Only The Beginning: preface

Music is as ancient as humanity. Its melodies and rhythms have not been preserved in the fossil record, but music is found in some form in all cultures; vocal chants and lullabies likely floated in the air for eons, well before the 35,000 year old bone flutes discovered by archeologists were carved. Neanderthals fashioned instruments from cave bear bones even earlier, 50,000 years ago. Music is a part of our DNA.

This book is not about ancient mythology, but the story of Orpheus from the 6th century BCE points us toward a variety of questions about how our brains are affected by listening and making music with and for others. A brief retelling of the myth precedes our story.

Legend has it that Orpheus was born of the Greek sun-god Apollo and one of the Muses, Calliope. Apollo invented the lyre, and was said to be a mean 7-string player. But when he gave his son his own lyre (made of either a tortoise shell or gold—his signature-model has been lost to antiquity), the boy Orpheus quickly exceeded his father's talents and was soon charming the birds and animals, and even the rocks. This original traveling musician joined Jason and the Argonauts on their Quest for the Golden Fleece, where he saved the expedition from the deadly song of the Sirens by playing and singing so beautifully that it drowned out the sounds that had lured generations of sailors to their deaths.

Upon his return, Orpheus met the incredibly beautiful nymph, Euridice. They were married soon after, but their marital bliss ended when his bride was bitten by a venomous snake and died. Orpheus' sorrow was so deep that no one could stand it, and his father advised him to go and rescue her from Hades, the underworld.

Orpheus put a fresh set of strings on his lyre, warmed up his voice, and found one of the few entrances to the Underworld. His melodies charmed Cerberus, the huge multi-headed dog that guarded the gates of Hades to prevent intruders or escapists. The sweet sounds of his lyre enticed Charon, the ferryman at the River Styx, to give him passage. And it is said that he played a killer riff that convinced Keith Richards (who, rumor has it, was manning the door) to allow him into the Underworld to find Euridice. Once inside, he sang and played a song of such heart-breaking longing that Hades (who had named the Underworld after himself, because he could) relented and promised to let Orpheus return with Euridice to the realm of the living, with one condition: His wife could follow him out, but Orpheus was not to turn and look back until he and Euridice were both completely under the sunshine in the world of the living. Orpheus agreed, and walked out with his wife behind him. But he worried about whether his wife was with him and turned around to check; she hadn't taken her final step into the light, so she disappeared forever back into the world of the dead.

Nothing Orpheus could do would open the gates of Hades again, and he pined away, refusing to look at another woman. That really upset the local ladies, so they tore him limb from limb and threw his dismembered head into the river with his lyre, where his head continued to sing a melody of heartbreak as it floated away. In memory, Orpheus had his own post-mortem cult and oracle. The name of this first master musician lives on in plays, movies, and album covers. Orpheus has served as subject and inspiration for composers ever since: Montiverdi's "L'Orpheo", Liszt's "Orpheus", Stravinsky's "Orpheus" ballet, and the wonderful Brazilian movie "Black Orpheus".

What can we learn from this musical myth? First, nothing brings a tear to the eye like music, even if you are the Boss of the Underworld. And the doubt that caused our hero to turn around suggests the hell of performance anxiety. Music can also help you sleep, put you in a trance, or bring peace and harmony to the world.

My own quest to explore how our brains create and respond to music comes from curiosity, not professional expertise. A career in veterinary medicine has given me a working knowledge of the nervous system and the sciences that study it, and an appreciation of the complexity of the living organism. As is true of many scientists, I have also been fascinated by music since childhood, although my abilities (born of persistence rather than talent) are very modest as an amateur guitarist/bassist/cellist/pianist/flutist/percussionist/and trumpeter.

In the spirit of Thoreau, "I should not talk so much about myself if there were anybody else whom I knew as well", I have included many personal observations. If I share too much of my own musical experiences, it is because I trust that other musicians and listeners have trod the same paths, and I am simply pointing out things that I have noticed along the way, Curiosity drives me to look thru the scientific literature, where researchers try to untangle musical myths from hidden truths. I have looked at music from every side, and I can't help wondering how it all works, how patterns of vibration in the air can make us laugh, cry, fall in love, and dance. Neuroscience, physics, psychology, and musicology offer some fascinating insights, but there is still much mystery and magic in music. Here are some things I have learned.

As Time Goes By: a tale of two cellos

John Holt and Ari Goldman had a lot in common. Both were accomplished contrarians in their own fields: John was an educator who believed that children were being stifled by rigid school systems and he wrote, spoke, and twisted arms to reform education. Ari was a writer and editor at the New York Times who became a professor of journalism at Columbia University, trying to instill the classic values of print journalism into students of the digital age who had never read a newspaper or learned the standards set by reporters of an older time.
John and Ari were also native New Yorkers who grew up with minimal involvement in music. But both men were struck with a passion to learn the cello in adulthood, both struggled with the challenges of learning an instrument when they were older, and both wrote poignant accounts of their musical journeys.

Learning music is good for you.
Why would anyone in their 40's, 50's, 60's, (or older) decide to take up the study of music? Most late starters report an inner flame of love for music which had never been consummated. Their parents couldn't, or wouldn't, give them music lessons or an instrument when they were young, and then education, career, marriage, and family took priority for time and resources. Some people say that they were never interested in music until a friend dragged them to a chamber music concert, or a vacation in Ireland kindled an obsession with Celtic fiddle music.

Many older music students point to a milestone that prompted a re-assessment of how they were spending their life, and they turned to music in the way that one might be reminded of a young love late in life, forgotten for decades but brought to mind by an old memory. For some, their children had left to start their own lives, while others might have lost their spouse and searched for something to fill an empty space in their days. A serious health problem may have reminded the person that it was now or never for that "someday list" that had always been hidden in the back of their mind. No matter what the reason, the need for a sense of purpose became a need for music.

One reason that late starters could offer to their spouses/partners to justify the hours away from family and the expense of pieces of wood and strings is that learning music is good for you. And, even though taking up pickleball or a class at the local community college would be a more practical and cost-effective way to maintain cognitive well-being, learning music, even at an older age, has demonstrable beneficial effects on the brain.

In a paper published in Frontiers In Psychology, Sofia Seifeld and collaborators compared 13 older adults who were given four months of piano lessons while 16 other seniors took painting lessons, computer instruction, or exercise. They found that cognitive function in the music group improved significantly more than it did in those who participated in the other activities. Learning piano improved executive function, inhibitory control, and attention. Visual scanning and motor ability improved, depression decreased, and positive mood increased. Overall quality of life scores jumped up as well. Seems like a pretty good package of cognitive benefits for four months of keyboard practice.

Results of a similar study reported in the European Journal of Neuroscience in 2023 pointed to one possible explanation for these findings. One-hundred and fifty-six people from 64 to 76 years of age who had never played music were enrolled in either piano lessons or a music-listening program for a year, and brain scans were done at 6 and 12 months. The piano players showed an increase in cortical thickness (that is, the amount of brain matter) and plasticity in five brain areas known to be involved in hearing, speech, and cognition. The improvements in coordination and cognition were bi-directional; that is, learning to move the fingers made the piano players smarter, and getting smarter improved manual dexterity.

In another study, the number of years of music practice, as well as the current number of hours spent playing music each week, corresponded with improved scores on executive function and spatial ability. The age of learning music didn't make a difference in this study.

Finally, music practice was found to improve the ability of seniors to understand spoken language amid loud background noise, which improves social function and delays dementia.

All of the evidence indicates that learning music is beneficial to the aging brain. But this should come with a warning label: It is hard.

John and Ari's stories of learning cello in adulthood are full of difficulty and frustration. Finding time to practice and working music lessons into a life full of career and family were more difficult than they had expected. Spouses and family members that would be encouraging and supportive if their 10-year-old was squeaking away on violin were simply annoyed when the 50-year-old worked on his cello; "Can't you take that in another room to practice?" has been heard more than once. Most people who take up a musical instrument later in life have already become accomplished in other areas of their lives, and struggling for a year to approach the competence level of a middle-schooler raises questions about how much ability the person actually had in other areas; maybe they were really an imposter, and their success in business or engineering was a fluke. And the older musical novice finds that there are new aches and pains to go with the normal aging complaints; "repetitive stress injury" rhymes with "practice" (well, not really, but you get the idea). Worst of all, the hard truth is that old brains and fingers don't have the learning ability of young brains and fingers.

Time for the confessions of a "serial late starter". I took up guitar when I was 14 and learned the basics easily. In college I played guitar in stage band and a series of rock groups, and kept up my playing with a contemporary choir at church, even though family and professional life had their demands. However, when my children left for college I decided that I could become a competent jazz pianist. I practiced several thousand hours in the evenings over the next five years, but the results were disappointing. For my 50th birthday I had a few friends over, played the handful of songs that I knew, and gave up piano, cold turkey. As W. C. Fields said, "If at first you don't succeed, try, try again. And then quit and don't be a damn fool about it."

After the piano attempt I took up the upright bass, a physically difficult instrument. Fortunately, the bar for competency on bass is pretty low and jazz ensembles always need someone to hold down the bottom, which allowed me to play regularly.

But then I moved from Seattle, with its vibrant amateur music scene, to San Diego, where amateur music was mostly jam bands and metal-doom-shoegaze rock bands—not for me. And it isn't much fun to play bass alone.

Searching to find my musical niche, I tried cello, violin, saxophone, flugelhorn, and a variety of percussion instruments in turn. With an understanding of music fundamentals and a knowledge of the instrument fingerings, any of those instruments should have been easy to learn. They weren't. It was like learning music all over again. Over and over again, with very mediocre results.

It seems like the scientific literature could help us understand why learning music is harder and slower as we age, but the information is scanty; apparently it is simply a given assumption that learning is easy for kids, hard for old people. Old dogs, and all that. Memory is the most obvious impediment experienced by the senior musician; learning is remembering, after all. Not only does the aging brain fail to retain and retrieve information as quickly, but the nerve pathways to and from the fingers slow down. Nerve conduction velocity, the speed with which a nerve impulse travels to and from the brain and fingers, declines with

age; in motor nerves it declines from 55.8 m/s to 52 m/s, and in sensory nerves it falls off from 59.4 m/s to 55 m/s. This might not seem like much, but a 5-10% slowing makes a big difference when you are trying to play 8 sixteenth-notes every second.
And it isn't just the brain and nerves that slow down. The aging body finds that arthritic fingers and stiff back muscles don't cooperate, even when they do receive the nerve signals. And sadly, playing music is less rewarding with age. When measured with the Barcelona Music Reward Questionnaire, there was a negative correlation between age and musical reward scores across the lifespan.
An optimistic Ari Goldman proposed that older learners can compensate for age with experience and trickery. Learning music involves problem-solving, an area in which experience and longevity offer an advantage. We just have to work harder at it.

From my own experience as a repeated late learner, and from the principles that brain studies provide, I have compiled my own list of "brain hacks" for older adults.

1. Exercise. Working up a sweat and elevating your heart rate improves circulation to the brain and stimulates the production of Brain-Derived Neurotrophic Factor (BDNF), a substance that brain scientists call "Miracle-Gro for the brain". BDNF prompts the brain to produce new neurons, something that previous dogma claimed was impossible after childhood. Run, hike, bike, or play pickleball, and then go home and work on that difficult Chopin piano piece.

2. Sleep. Learning occurs during the next sleep after your practice, when information rehearsed while awake is consolidated and disseminated across the brain. In an ideal world, each practice session should be followed by a good nap.
3. Trust your subconscious. Every experienced musician knows that musical skill progresses, not during practice, but a day or a week later. Short, frequent practice sessions give time for the inaccessible parts of the brain to sort out the notes and pathways for the fingers, and we have to have faith that what we practice will eventually show up in our playing..
4. Protect your hearing and have your hearing checked and corrected if necessary. This isn't just for old rockers who lost their hearing playing in front of a six foot stack of Marshall amps; orchestra musicians are exposed to volumes that are well above ear-damage levels. Even slight hearing loss impedes your ability to hear the nuances of the sounds that you produce; I had high frequency hearing deficits for 10 years before I agreed to get hearing aids, and I thought I was fine. This was during my attempt at learning violin, and after I got hearing aids I discovered that my tone was even higher and annoyingly squeaky than I had thought. I had to adjust my technique to eliminate the screechy sounds that I didn't even know I was producing.
5. Use "deliberate practice". The common thread with expert musicians is that they use their precious practice time concentrating on challenging material and specific skills,

rather than repeatedly playing thru music that they already have under their fingers. On the other hand, life is short, so don't forget to play and enjoy music that you know well.

6. Use mental practice when precious practice time is scarce. As you lay in bed trying unsuccessfully to fall asleep, picture the fingering for your scales or review that four bars of a piano piece that you have been trying to memorize. While you are on a long boring drive, imagine the shapes of the guitar chords that you been working on. Mental practice away from the instrument has been shown to be almost as effective as practicing with instrument in hand, and reduces the repetitive stress of physically playing a passage over and over.
7. Slow down. There is a saying that "to play fast, practice slow". Our fingers can easily move as fast as a heavy metal guitar shredder, but we are limited by how quickly our brains move between the notes. Fingering patterns soak into the brain more effectively when played repetitively in slow motion. The more times something is played slowly, the easier it is to speed up. After your short nap, that is.
8. Finally, play with other musicians. Jazz musicians refer to technical ability as their "chops" (partly because horn players are using their cheek muscles, but also because extended concentrated practice is referred to as "woodshedding", a throwback to when the guitar player would spend hours practicing out behind the barn). There are "practice chops", "rehearsal chops", and "gig chops", and they are different. It is a common experience that a

novice player masters a tune in practice, but as soon as they play it in rehearsal, it falls apart. Once they can play it smoothly in a practice session and have an opportunity to play in front of an audience, they find that the same music has left the building. Find a chance to play with others and to play in public.

If this all sounds like a lot of work, it is. As beneficial as music is for the brain, if learning to play the piano or guitar is like taking medicine, it isn't worth it. But for many, music of any style provides an opportunity that may not have been available in younger years; for those latent musicians, the quest to master an instrument is worth the cost.

Like many adults, John Holt was not exposed to much music growing up in his home in New York. His college years coincided with the heart of the Big Band era, and when he discovered Benny Goodman and Woody Herman he became a passionate music listener. He then expanded his interests to the music of Stravinsky, Brahms, and Mendellson.
Holt's sister suggested that John teach elementary school, even though he didn't have a teaching degree. While teaching at various private schools he became disillusioned about how children were taught, and wrote "How Children Fail", a scathing but influential book about how schools were discouraging kids from learning. Children, he believed, would learn naturally on their own if allowed to explore their interests with abundant resources to draw from. His book made Holt an iconoclastic guru

for those who felt that children could learn without being taught, and the homeschooling movement was launched. Now famous, he published "How Children Learn" and eleven other books about learning, including "Never Too Late; ", his memoir of learning music as an adult.

With his musical passion ignited by big band jazz and orchestral classical music, John dabbled with guitar in his 20s and flute in his 30s, but being largely self-taught (which fit well with his educational philosophy) and short of time and money, his playing was only an occasional amusement.

At 40 years of age, Holt impulsively bought a cello from one of the students that he was tutoring, and for a while he made serious efforts to learn to read and play chamber music. After he had been playing for only a year, another student asked Holt to teach him how to play; John could only protest that he had only been playing for a year himself, but he followed the adage that "teaching others is how adults learn", and both of them benefited.

Time and the demands of his new lecturing career slowed his progress, but when he was 50 he rededicated himself to becoming an accomplished cellist. John studied with a teacher, but ultimately discovered that he grew more by simply trying to play music with others. Becoming obsessed with his cello journey, Holt attended summer music camps and joined community orchestras, always surrounded by players much better than himself.

Ever the educational theorist, Holt offers a rich collection of ideas about learning music as an adult.

"As long as I pay attention to what I am doing and hearing, whatever I did well, I will tend to do again."
"But we don't and never did learn complicated skills by dividing them up into a series of tiny steps, each of which we had to get right before we could go on to the next.
"The beginner needs other people to play with."

At the time of the writing of "Never Too Late", John Holt played cello with a community orchestra in Cambridge, Massachusetts on Mondays, joined a quartet to play Mozart on Tuesdays, and practiced a Mendelssohn trio with violinist and pianist friends on Fridays. And every day his cello would rest in the corner of his office at school, inviting him to steal a few minutes for practice at lunch.

Ari Goldman had also not grown up with music, but developed an interest for classical music in middle age. He struggled with all of the demands of his career as a journalist and professor at Columbia University, and experienced all of the frustration of learning with a brain tuned to writing and speaking, not playing the cello. He spent time with teachers, both supportive and harsh. When his young son took up the cello and quickly surpassed his ability, they both attended week-long music camps; he was full of pride for his son and embarrassed by his own limited ability. He joined the Late Starter Orchestra in New York City, where he could hide his meager skills by only playing the notes that he could find as the orchestra made their way thru the Brandenburg Concerto. With his 60th birthday looming, Ari set himself a goal of preparing pieces to play for his birthday gathering, but his teacher dropped him, saying that the wasn't progressing. A despondent Ari heard the ghostly voice of his first teacher and mentor urging him to go back to playing only open strings, trying to hear the simple beauty of a horsehair bow pulling gently across the C, G, D, and A strings. After a month of reconnecting with the beautiful voice of his instrument he returned to his Bach, and when the time came to play for the 70 people gathered for his birthday, the music took over and Ari was lost in the centuries-old melody, the warm feel of wood and strings, the neural synapses that seemed to have developed without his awareness, and the appreciative applause of his family, friends, and colleagues. His musical memoire, "The Late Starter Orchestra", is an inspiration to anyone who thinks that their life still has more possibilities.

I tried to find a suitable conclusion for this essay, but John Holt sums it all up better than I could: "The point is important; to take up music might help many people solve, or better bear, their troubles—but not if they take it up only for that reason. To get any help or health from music (and I suspect, anything else), we must come to it for its own sake."

It Don't Mean A Thing: the secret of swing

A bright San Diego sun warmed the green grass where Scripps Hospital was holding their staff picnic on this Sunday afternoon. Although I didn't work for Scripps (I was a veterinarian doing house-call practice), I had been invited to play bass in a jazz trio that provided music for the occasion. The drummer was Gary, a pediatrician with whom I had played music off and on, and the pianist/singer was Peter, a rheumatologist who was a stranger to us. Peter suggested that our extemporaneous group of medical folks might be called Three Doc Night.

As we worked our way thru a list of standards from the Great American Songbook, I couldn't tell what was different as the songs of George Gershwin, Irving Berlin, and Cole Porter came to life with more verve than usual. I guessed that somehow the pianist's timing provided a spark to the music, yet I couldn't hear how he did it. We all sensed that our amateur trio shared a special feeling as we played over the smell of hotdogs and the voices of the picnickers. We were swinging.

The concept of swing has always been controversial and difficult to define, for both musicians and listeners. Musicians claim that you can feel it, but you can't explain it; there must be some secret to swinging that even musicians themselves don't understand. Researchers with oscilloscopic analyzers and computer-altered recordings have tried to analyze why some music swings, and some doesn't; some of their findings make sense and some don't.

On a macro-rhythmic level, any musician can tell you that the term “swing feeling” suggests unequal time durations within pairs of eighth notes. In standard musical notation, a pair of eighth notes (whether actually played or merely implied in the melody) divides each quarter note beat evenly in half, da da, da da, da da, da da. But if the first note of each pair steals a little time from the following note, the second note is decreased proportionately, daa da, daa da, daa da, daa da.

The ratio of eighth note durations in “straight time” could be expressed as 1:1, each note dividing the beat neatly in half. In the most common version of “swing eighths” the note durations have a ratio of 2:1. This could be indicated in written music by writing each beat as a triplet of three eighth notes, with the first two in each triplet tied together to make one. The resulting thicket of extra notes and ties uses too much ink, so if the music is marked “swing”, the assumption is that all eighth notes will be swung with a 2:1 ratio unless otherwise noted.This explanation is the first thing that student jazz band members will be told.

If unequal eighth notes distinguish “swing rhythm” from “straight time”, there remains a broad range of interpretation. One musicologist created a visual representation which he called a “swing-ogram”, and found that the swing ratio found in the ride cymbal rhythm of jazz recordings varied from 1.48:1 to 1.82:1. But when musicians were asked to play on computer-controlled MIDI keyboards or wind instruments, the ratio was more like 2.45:1, or even up to 3.5:1 at very slow tempos.

There are many subtle shades of swinging eighth notes. During the bebop era Charlie Parker, Dizzy Gillespie, and other adventurous musicians started playing tunes that were melodically jagged, harmonically complex—and fast. Virtuoso horn players attacked the chord changes with so many notes in such short periods of time that the 2:1 ratio of Glenn Miller's "In The Mood" became more like 1.5:1 for Charlie Parker's "Ornithology"; still swinging, but the toe-tapping that it inspired in the audience was less noticeable.

At the same time that bebop was changing jazz, the blues were moving out of the Deep South into Chicago and starting to cross the racial lines that had kept Black music in its place. Part of the appeal of the blues was the obvious use of uneven eighth notes. As "Goin' To Kansas City" blares from the juke box, the "shuffle rhythm" of unequal eighths pulls couples out of their seats and onto the dance floor. Even the casual listener can hear the daa da, daa da of a "deep swing" feeling where the first note of each pair seems to claim even more of the time, a ratio of 2.5:1. For an even deeper swing, imagine music that might be used as a stripper takes it all off.

The conventional explanation holds that the secret of swing lies in the uneven durations of the eighth notes, but the "swing ratio" fails to explain why some music generates a strong feeling of swing, and some doesn't. It seems that there is a deeper secret that even the musicians can't explain.

Enter the scientists. Musically-inclined researchers have hypothesized that that swinging also has a lot to do with "micro-timing"—tiny deviations from the beat that can't be consciously perceived by the listener. Musicians have long argued that the feeling depends on slight delays in each downbeat, but the evidence is confusing.

Some studies have manipulated jazz recordings by artificially "quantizing" the music so that the notes fall exactly on the beat. Listeners were asked to rate how much each version swung; the music that was altered to eliminate any slight delays in the downbeat were rated as having more swing, despite what jazz musicians claim. These researchers concluded that "our results suggest that naturally fluctuating MTDs (micro-timing delays) are not an essential factor for the feeling of swing."

Mathew Davies authored a Portuguese study using synthesized rhythm patterns for jazz, funk, and samba, varying the timing delays of downbeats from none to exaggerated. Listeners rated the versions with no timing delay as higher in groove, liking, and naturalness, with expert musicians showing even more preference for the exact rhythms than untrained listeners. One has to wonder if the results would have been different if the listeners were from Philadelphia rather than Portugal.

The findings of this musical research seem to depend on the methods used for analysis. A different conclusion was reached by researchers at the Max Planck Institute for Dynamics and Self Organization in Germany (which sounds like it would be a mind-altering place to visit). There, physicist Theo Geisel (no, not the children's author of the same name) analyzed 456 jazz piano recordings and found that slight downbeat delays increased the feeling of swing, according to his audience of musicians. In his paper in the scholarly journal Communication Physics, jazz musicians were nearly 7.5 times as likely to rate music as swinging when the soloist's timing was partially delayed with respect to the rhythm section,

Geisel's more nuanced analysis showed that the swinging soloist delayed the downbeat of each eighth-note pair by 30 milliseconds (the shortest time that even a trained musician could perceive), or 9% of the duration of the note, compared to the rhythm section. The shorter second note of each pair was played in sync with rest of the band. This finding seemed to support the "Participatory Discrepancy" theory of Charles Keil, who claimed that the first note, slightly delayed, creates tension in the listener's mind which is resolved when the upbeat arrives in sync with the rhythm section. The result is greater "entrainment", an urge to move to the beat. One of the studies quantified the need to bop to the music with a camera that captured the listeners' head movements as the music played.

Other studies generally agree that the secret of swing is hidden in slight downbeat delays followed by synchronization of the offbeat, but how does this relate to the playing of the jazz greats? In the Journal of New Music Research, 2018, Christian Ditmar looked at three classic jazz recordings to see if there was a pattern.

On "The Red Door", baritone saxophonist Gerry Mulligan is backed up by legendary drummer Mel Lewis. On this recording, Mulligan's downbeats were placed squarely with the ride cymbal beat, but his offbeats were not well-synced with the drummer. Micro-timing delays on the downbeat were not apparent.

When Dittmar looked at "Down Under", a track by Art Blakey's Jazz Messengers, he found that trumpeter Freddie Hubbard laid

way back on the downbeats, but synced to Blakey's offbeats. Score one for Participatory Discrepancy.
"Blue Trane" is a classic track from John Coltrane's early group, driven by Philly Jo Jones on drums. An analysis of trombonist Curtis Fuller's solo, which swings just fine, found that neither the downbeats or offbeats were well synced with the drummer's ride cymbal beat.

Every musician seems to have their own feeling for time; when the whole band hears it the same way, there is an indefinable sense of propulsion that entrains the body and soul. When the time isn't happening, it is hard to tell why, but the urge to move and groove is gone. During the 1950's the East Coast jazz scene, which rested on an urban African American background, often struggled to feel the swing with players from the West Coast, who were more often white and from a laid back musical environment. Latin musicians, who grew up surrounded by clave rhythms and straight-time patterns, brought more rhythmic spice, but a different sense of where exactly the notes sat in time. And most young players today grow up listening to rock and pop, and struggle to capture the subtle rhythmic nuances that make jazz sound like jazz.

Playing fast and loose with the timing of eighth notes and shaving a couple of milliseconds off the downbeat are not strictly degenerate trends of the 20th century; other musical genres through the centuries have used flexible timing to encourage their audiences to move.

Even classical music has its own history of note inequality. During the Baroque Period in particular, music heard in the elegant court ballrooms often used unequal note rhythms. Scores of Baroque compositions that are written for modern day orchestras are full of dotted eighth/sixteenth note combinations, which should have a precise ratio of 3:1 if played "correctly". In their day, however, the unequal eighths were more likely played with a ratio of 2:1. We have no way of knowing what Handel's Water Music actually sounded like at the time, but musicologists suggest that the written notes may have actually been performed with a swing feeling, and the court dances may not have been as stuffy as they are portrayed. Many early-period musical ensembles try to recreate older music with a more flexible interpretation that we might call swing

Micro-timing also puts life into the Celtic country dances. During the Covid pandemic I took violin lessons online from Riley, my granddaughter's teacher, and tried to capture the jaunty feeling of Irish fiddle tunes. I found it difficult to create the lilting triple-time feel of the jig until my teacher revealed the secret to playing Celtic dance music: Jigs consist mainly of groups of three notes, played at a lively tempo, alternating down-bow and up-bow strokes. Riley suggested that even when playing three (theoretically) equally spaced notes, I should use the entire length of the bow for the first note in each group, and half the bow for each of the succeeding two notes. Even without trying to give the first note extra time, using an unequal length of bow forced that effect; I couldn't actually hear a timing difference, but the feel was suddenly there.

The waltzes of old Vienna often use similar micro-timing tricks to add a little extra swoop to each first step. Note inequality and imperceptible delays are also used in some forms of Middle Eastern dance music. It seems that we can unconsciously feel even tiny differences in timing, and musicians have been using some form of unequal timing for centuries.

Swing, both in the form of unequal eighth notes and of imperceptible rhythmic discrepancies between the instruments, creates a sense of movement, our bodies trying to resolve the minute variations in the flow of time. Swing creates tension in the listener, but the variety of applications suggests that it also identifies the human element in the music. When musicologists studied the effects of swing by artificially manipulating the timing of recordings or creating computer-generated music, listeners found it less pleasurable, but when the variations occurred naturally in a performance, rhythmic variability made all the difference. When real human beings are playing, the swing ratio varies, increasing at slow tempos and decreasing in faster tunes. Often a jazz soloist will decrease the depth of swing at the end of a phrase, and even more to suggest that their solo is coming to a close. Manipulation of the swing feeling may be one of the ways that musicians seem to communicate with their bandmates without giving any visible signs, as if reading each others' minds. Slight downbeat delays that sync on the upbeat serve to create an invisible bond between the musicians, and between those who create the music and those who listen. It isn't surprising that the term "time feel" is used to describe the effects.

It's all about the feel, and swing reveals the human element in music.

As our jazz trio dug into "There Will Never Be Another You" at the hospital staff picnic, we were swinging and the music seemed to carry us along effortlessly. I couldn't tell what it was that pianist Peter had, but it was contagious; my own walking bass lines had never been more swinging. The three of us anticipated the "three-tag" ending as Peter played the final bars, without having planned the ending in advance. As the last notes faded, we looked at each other and grinned.

Shake It Off: music performance anxiety

As the lights in the great hall dimmed, an ornate velvet curtain parted to reveal an imposing black grand piano at the center of the stage. The audience waited in anticipation; the concert had been scheduled to commence ten minutes ago, but the only sound that could be heard in the silent auditorium was a faint scuffling noise from backstage. Finally a balding man in a tuxedo entered the stage, wary as a deer entering an open meadow at twilight. After seating himself on the piano bench, there was a pause while he took a few deep breaths. Raising his hands to the keyboard, the opening notes of Liszt's "Piano Sonata in B Minor" floated out over the audience, enchanting music from one of the greatest classical pianists of all time. But Vladimir Horowitz had to overcome a wave of panic every time he walked onto the stage. Sometimes, it is said, attendants had to drag him bodily from his dressing room and push him into the spotlight.

Horowitz is celebrated for his interpretations of music from the Romantic era, but he is just as famous for his struggle with Music Performance Anxiety, or MPA (I know, it sounds like a craft Belgian beer, but we will use this abbreviation for convenience). He tried all manner of rituals to calm his anxiety when he played in public, but had to discontinue his concert career for twelve years due to his crippling fear of performing.

Musicians of all ages and abilities spend their lives learning to play an instrument or sing so that they can share the gift of music with listeners, so it seems strange when their hands tremble, their eyesight gets blurry, and their hearts race as if they are being chased by a lion (I don't know why, but neurologists always use "chased by a lion" to illustrate the physiologic hallmarks of fear). The great Russian pianist was not at all unusual in his struggles with MPA; 30% of all classical musicians report varying degrees of anxiety, varying from uncomfortable inconvenience to career-ending debility. No discussion of "stage fright" is complete without a list of sufferers, and music lovers will easily recognize these famous musicians by single names alone: Streisand, McCartney, Gould, Chopin, Van Halen, Pavarotti, Simon, Rihanna, Adele, Bocelli, Cher…the list could go on forever. MPA is an equal-opportunity demon; the nine-year-old at her first piano recital or the thirty-year-old housewife auditioning for a part in a community musical is just as likely to freeze and pray for the floor to swallow them up as is the famous virtuoso whose entire career is on the line. Women suffer the condition more often than men, and classical musicians report fear of playing for an audience more than performers in other genres, but that may be because both of these groups experience more harsh judgement than the male rock star, who may also be generously self-medicated before taking the stage.

A thoughtful and authentic first-hand account of Music Performance Anxiety can be found in Sara Solovitch's charming-but-terrifying book, "Playing Scared: A History and Memoir of Stage Fright". Her parents and teachers all agreed that Sara was very musical, and she played the piano well at 10 years of age. But when she had to play for an audience, whether at a piano class recital or at a music competition, she invariably suffered a total meltdown, sweating and shaking so much that she could barely control her fingers on the keys. Her parents were proud of her ability, but embarrassed by her obvious emotional distress. Sara continued to study music and attended the Eastman School of Music's preparatory program, but continued to be plagued by MPA; when it came time to choose between attending a music conservatory or an academic university, the choice was easy: She gave up music and the gripping fear that accompanied every performance. But that wasn't the end of her story.

Sara always owned a piano, but would only revisit a few of her favorite pieces when no one was listening. After raising two boys who were confident and musically accomplished, she remembered how much she had loved the piano, found a teacher, and practiced in private for six years; she was still too scared to play when anyone was listening. Sara undertook an odyssey to learn everything she could about stage fright so that she could conquer her demons and find pleasure in sharing her music with others.

Anxiety disorders are so pervasive, and so stubborn, that one can only wonder why such a disabling trait should come standard in our mental toolkit. Leaving aside the actual being-chased-by-a-lion scenario of our distant ancestors, it seems we would be better off if we could remain cool, calm, and collected, even in the most stressful of circumstances.

As a veterinarian with a special interest in animal behavior, fear and anxiety are some of the most common behaviors that call for my help, especially in dogs. I offer my clients a vastly over-

simplified explanation from brain function and anatomy, which applies to humans as well.
The amygdala is a small almond-shaped structure deep in the temporal lobe of the brain, famous for being the seat of emotions. It reacts to sensory inputs from the thalamus and memories from the hippocampus, and responds with messages of aggression, sadness, or fear and anxiety. The amygdala instantly notifies the adrenal glands, which produce adrenaline and cortisol, those chased-by-a-lion substances. Then, just to make sure, the amygdala sends a memo to the frontal cortex of the brain, where the situation undergoes further analysis: Are we walking across a wild savanna in Africa and that lion's roar means impending doom? Or are we at the zoo with a sturdy twelve-foot fence between us and the King of Beasts? If it is the latter, the frontal cortex sends back the message that there is no danger, and the initial surge of stress hormones subsides, leaving only the invigorating feeling of a movie jump-scare.
The important point of the fright reaction is that by the time the frontal cortex analyzes the threat 20 milliseconds later and sends an all-clear message, the cascade of stress has already sent the heart racing, raised the blood pressure, and induced secretions and excretions from all parts of the body. The reassuring memos from the executive parts of the brain are too little, too late; the body already feels like something bad and dangerous is in progress.
This physiologic explanation explains the sequence of fear events in the body, but it doesn't explain why the amygdala triggers the

same chased-by-a-lion reaction when a person plays Mozart's "Piano Sonata No. 16" in front of a dozen listeners.

Music Performance Anxiety is often considered a type of social phobia; we are built to care deeply about what others think of us. Reaching back to our ancestors in their small hunter-gatherer tribes, it really does matter whether others in our group accept us or reject us. Disapproval could be followed by expulsion, which threatened the person's very existence. It isn't surprising that sufferers from MPA often have other social phobia tendencies. Still, hitting a few clunkers among a thousand notes in a sonata or forgetting some of the words in a popular tune in front of an audience doesn't really seem like it should get you kicked out of the clan.

Perfectionism is often blamed as an underlying cause of MPA, and many musicians with stage fright tell stories about the strictness of their music teacher. In the old days, the intimidating gray-haired violin teacher might stop their student if they played a note that was out of tune and yell at them in Russian to correct their mistake. Maybe being chased by a lion would have been preferable. But in our enlightened era, teachers are supposed to be supportive, responding with loving encouragement to their students' barely discernible rendition of "Twinkle, Twinkle, Little Star". And yet, anxiety seems to be a epidemic.

Not all of the blame should be placed on teachers or "tiger mom" parents who pressure their children in the hope that it will result in greatness. Shakespeare may have been thinking of stage fright when he wrote that "The fault lies not in our stars, but in ourselves." While 30% of orchestra members are affected by

MPA, 70% are not, and even the ranting of an overbearing conductor doesn't make them break a sweat. Leaving aside our different cultural upbringings and family dynamics, we are all wired a little differently. Perhaps the lucky 70% are blessed with a less-reactive amygdala, which sends out its danger message in small print, rather than bold capitals (although this might be a handicap when the music calls for more emotion). The analytic frontal cortex may be more persuasive in some individuals, overriding the panic signals from the amygdala. It has been proposed that MPA sufferers have a greater number of beta-adrenergic receptors on their cells, which causes them to respond to the adrenalin rush in a more powerful way. To some extent we can't change how we are made.

Some people have the advantage of being less sensitive to the judgements of those around them, but not everyone is so fortunate. MPA is serious business to those who feel like vomiting at the thought of performing. Performance psychologists are in demand in many arenas: In sport, the "yips" can cost a professional athlete their career, or an Olympic athlete a spot on the podium. Public speaking is part of the job of many corporate execs, but the CEO that perspires profusely on the podium and speaks so fast as to be unintelligible will have to find someone else to deliver their message. MPA is a dirty little secret among many musicians, carrying the stigma that perhaps they lack talent or don't practice enough. Increasingly these people are looking for solutions. There are many answers, but no easy fixes (with one partial exception).

Talk is cheap (well, depending on the therapist), and some form of Cognitive Behavioral Therapy is the first thing recommended to cure MPA. Tell yourself that the audience are your friends and want you to succeed. Or picture them in their underwear, as nervous public speakers are advised to do. Or imagine that you are playing for farm animals. Talk yourself out of your perfectionist mindset. When I attended a jazz summer camp full of talented teenagers and old wannabes like me, John Clayton Jr was the mentor for the bass players. Clayton, a tall world-class bassist and one of the kindest people you could imagine, reminded us not to take the music too seriously: "Nobody ever died because you played a wrong note". Unfortunately, most of these reassurances fail to calm the nervous.

Your music teacher might tell you that you need to practice more, that fear of failure can be overcome by having the music perfect before you perform it. This can backfire, however; music is never perfect, and endless hours in the practice room may simply build up a barricade of remembered mistakes to dread during performance. Be ready, but not too ready.

Just relax. Some MPA suffers have found that meditation, yoga, or even hypnosis helps. At least do some deep breathing exercises before walking on stage.

Some amateur neuroscience nerds relate MPA to the division of the right and left brain. According to the concept of hemispheric dominance (which is true, but over-simplified), the right brain is artistic and wholistic, concerned with creativity and positive emotions, while the left brain is the editor, criticizing the output of the right brain and pointing out each mistake and flaw.

Although the two halves of the cerebral cortex communicate with each other, the accountants in the left brain (where language lives) tend to overshadow the free-and-easy hippies in the right brain. One performer suggests that the anxious musician clench their left fist tightly (which is controlled by the right brain) before taking the stage, to pump up the power of the right hemisphere.

If making a fist doesn't work, there are always drugs. Shocking, but musicians have often resorted to self-medicating with a pharmacopia of strong chemicals. Heroin and marijuana were traditional for jazz musicians in days gone by when they needed something to "take the edge off". Rock musicians expanded the menu to psychoactives, including cocaine and amphetamines, to go with the pot and alcohol that were already plentiful in club venues. Valium and other benzodiazepines act more directly on anxiety and have found use by many pop performers. Unfortunately those drugs also interfere with memory, which doesn't help when a singer forgets their lyrics. Classical music is more technically demanding, making psychoactive drugs and alcohol potentially hazardous. Still, the concert pianist or violin soloist whose hands shake and sweat during performance may need a little chemical help.

Propranolol to the rescue.

In the 1960's a Scottish pharmacologist, Sir James Black, discovered beta blockers. Adrenaline creates its many effects by binding to receptors on a variety of cells in the body via the sympathetic (fight, flight, or freeze) nervous system, and "adrenergic receptors" come in two flavors. Alpha receptors cause constriction of blood vessels and small airways in the lungs, as well as contracting the bladder and decreasing the response of excitatory nerves in parts of the brain. Beta receptors increase heart rate, constrict blood vessels to raise blood pressure, trigger perspiration, and generally prepare the body for an emergency. Sir James found that certain drugs could block the effects of adrenaline on the beta receptors, particularly in the heart and blood vessels, with life-saving effects in heart disease and hypertension. He received the Nobel Prize in 1988, and beta blockers became one of the most prescribed medications world-wide.

In 1976 an article in the British medical journal, Lancet, suggested that blocking the beta receptors might help with the bodily symptoms (racing heart, trembling, sweating) of the anxiety that plagues public speakers, actors, and musicians. A decade later a survey revealed that almost a third of classical musicians used the most common beta blocker drug, propranolol. Despite widespread use of beta blockers to treat MPA, musicians are split on whether they should be used, or even allowed (although there is no anti-doping authority in music; they would have more important issues than misuse of blood pressure medication). Many players insist that a little bit of adrenaline-fueled nervousness is needed to produce performances filled with emotion and excitement, and feel that propranolol would make their playing lifeless. But to the MPA sufferer the racing heart, trembling, and sweaty palms are a greater impediment. When conservatory music students were given beta-blockers before being judged by faculty members, the performances were rated more highly. Although most don't want to admit it, for many accomplished musicians beta blocker medication is just as necessary as a violinist's bow rosin or a guitarist's pick. Clearly blocking the effects of adrenalin can ease performance anxiety; but does it is help recovery from MPA?

The brain tells the body what to do and how to feel, but it works the other way around as well. The concept of "embodied cognition" was introduced by William James, the "Father of Psychology", in the late 19th century. He claimed that the bodily feelings of fear, racing heart, and knots in the intestine were responsible for creating the brain's experience of fear: "When we see a bear we don't run because we are afraid, we are afraid because we run." This concept has been further developed using modern scientific techniques by neurologist Antonio Damasio, whose book "The Feeling of What Happens" shows how the brain listens to what the internal organs are feeling in order to know what it should be thinking. The use of beta blockers can change the messages that the body sends the brain, essentially saying "What are you worried about? If there was actually danger you would be feeling it." Over time the brain can learn to come off high alert and remain calm, even without the drugs.

In the canine patients in my veterinary practice I will often use beta blockers along with behavioral therapy to treat chronic anxiety problems, since they are relatively benign medications (compared to the tranquilizers and mood drugs). Beta blockers seem to be most useful in those dogs with measurable adrenaline over-stimulation—high heart rate, pupil dilation, and increased blood pressure. I can't tell whether it improves their performance at the piano, however.

One of the cornerstones of treating anxiety, in dogs, rats, or humans, is desensitization therapy. If the animal/person is exposed to the anxiety trigger repeatedly without experiencing anxiety, the stimulus gradually loses its power. We now know that memories, good and bad, work like files on a computer hard drive: When a musician looks out over an audience, the "file" of previous stage fright experiences is opened and the person is reminded of previous panic reactions. But like a file on your word processor, the file can be changed and the new version saved. If fear is not experienced during the experience, the new version of the memory is shorn of some of its emotional content. By changing the fearful nature of an episode, the memory with less emotion replaces the fear-laden original. The memory may not become pleasant, but at least it becomes tolerable with repeated exposures; rinse and repeat, for a hundred repetitions. Unfortunately, exceeding the person's tolerance can start the whole process over again. The musician who loads up on propranolol for an important concert but hasn't spent a year gradually desensitizing themselves in low-stakes situations is likely to have a meltdown and then will have to start desensitizing all over.

Sara Solovich had suffered from MPA ever since childhood, but she loved the piano. After abandoning any hope of a music career during her university years, she pushed music to the side. When she turned 49, she had the urge to return to her passion, but she knew that performance anxiety was a dragon that she had to slay in order to play. Sara embarked on an epic journey to understand and confront the fear that cripples so many musicians, public speakers, actors, and athletes. She interviewed every expert in the field, talked to scores of musicians, and asked the help of piano teachers with a wide range of philosophies.

Finally, after ten years spent exploring the causes and cures for MPA, she knew that she would have to prove to herself that her anxiety could be overcome. Sara committed to a goal of playing a recital for friends, family, and well-wishers for her 60th birthday. There is nothing like a deadline to sharpen your focus and increase your willingness to try anything that might help. She had less than a year to find the secrets that would set her free.

She tried yoga, meditation, and Cognitive Behavior Therapy, but still felt the old familiar nerves.

Working with sports psychologist Don Greene and her piano teacher, she embarked on a program of gradual desensitization, And lots of propranolol. With medication to still her racing heart and sweaty fingers, she played for one or two friends at her house. She took master classes and small recitals with her teacher (attended by 10 year old students and their adoring parents). She went to a piano camp, where she almost managed to make her way thru a performance of one of her familiar pieces. Her teacher took her to an airport, where there was a piano in each wing of the terminal, and she joined other students in running thru her pieces as uninterested passengers passed by on their way to their flights.

On the day before her 60th birthday, the moment had come. The hall filled with friends, family, and supporters. Sara took her beta blockers, walked to the grand piano, sat down, took several deep breaths, and mentally centered herself. As an opener she played a Haydn duet with her son Max, an accomplished violinist. Then it was time to face her fears. Her fingers reached for the notes of Debussy's Reflets dans l'eau (Reflections in the Water), and the notes rippled forth, as if they had a life of their own. Any minor mistakes were washed away, like a stick disappearing in the river as it flowed on. It had taken most of a lifetime, but her self-consciousness ebbed away and she could actually enjoy the applause of the supportive audience. She felt like she was standing alongside the music, hearing the beauty without wondering what the listeners thought of her. This was what playing music should feel like.

Both Sides Now: handedness in music

Pick up a pen and write your name. You will probably take the pen in your right hand and scribble your signature without thinking about it. If you pick it up in your other hand, however, you will have to concentrate hard even to produce script worthy of a first-grader.

If you are a lefty born before 1950, chances are that your grade school teacher instructed, cajoled, and threatened you to line up your paper in the same direction as your classmates and put the pencil in your right hand. Generations were beaten into submission to write like the righties, until eventually most teachers gave in and allowed the lefties to line up their paper and pen in a mirror image of their classmates. There was still the problem of dragging the heel of their writing hand thru the fresh ink of what they had just written, but it was worth it.

Handedness has been a popular subject for scientific study. Some sports-loving neuroscientists spend their evenings poring over century-old baseball statistics, comparing obituary records to find out if right-handed or left-handed players lived longer. Psychologists study whether handedness has an influence on creativity or mental illness (yes on both, to an extent). Generally about 90% of the human race seems to favor their right hand, suggesting that their left cerebral cortex is dominant over the right, since most functions of the nervous system cross over. It would be interesting to find out what proportion of handedness researchers favor their left hand; lefties often seem more concerned with the challenges and unfairness of a right-handed world than the other 90% of their colleagues.

If dexterity (from the latin for right handed) favors the righty, what are the implications for musicians? Most musical instruments are designed for people to put their best hand (ie, right) forward, and it seems that the people who are sinister (the Latin for left-handed) would be at a distinct disadvantage.

Perhaps, the lefty neuroscientist wonders, the reason he was unable to master Chopin's Nocturnes as a piano student was because he lacked hemispheric dexterity (although Chopin himself was left-handed).

Still, intuition suggests that a lefty in a right-handed world would suffer obstacles when using a right-handed instrument to play right-handed music. (Consider a typical two-handed piano piece, with the right hand playing beautiful melodies and rapid arpeggios, while the left hand is relegated to occasional bass notes and chords.)

Are right-handers or left-handers more common among musicians? Does the dominant hand determine the skill level of the music student or the concert virtuoso? As is often the case in scientific studies, the results are mixed.

A 2023 article in the journal Music Perception found no difference in hand preference between grade schoolers judged musical and those with no musical ability (although the righties were academically more successful). Similarly, a European analysis of musicians in seven different orchestras found the proportion of handedness was similar to the general population.

On the other hand (so to speak), a 1994 study in Music

Psychology found that more professional orchestra musicians were left-handed than in the general population (12% compared to 8%). Perhaps this was not due to some musical talent conferred by using the non-standard hand, but by the fact that left-handers are in general more attracted to creative and artistic pursuits.

Which brings us to a broader question: Why would humans develop an asymmetrical preference for manual tasks? It isn't really the hands, of course, it's the brain. The right hemisphere of the cerebral cortex controls the left side of the body, and vice versa.

Until very recently, it was claimed that humans were the only species in which the two sides of the brain differed in their specialization, and this was attributed to the development of language. The ability to communicate (and even think) in words and grammar is housed in the left side of the cortex, and the idea was that the right side of the evolving brain could take care of the usual daily activities, freeing the left for the communication experiment that ultimately resulted in language. This grand assumption of human superiority has faded, however, as hemispheric dominance and right/left handedness have been found to be widespread in the animal kingdom.

One of the earliest observations to challenge the humans-are-the-only claim came from a colony of monkeys in Japan. The primates lived on a small island surrounded by a moat, which kept the monkeys confined because they hated to get wet. In everyday activities, such as feeding and grooming, the animals showed no preference which hand (or foot, for that matter) to use. But their caretakers noticed that when they threw oranges (the monkey's favorite treat) onto the island, a few would always end up in the moat. The monkeys wouldn't venture into the water, but they figured out that they could lean out far enough to grab an orange if they held onto a branch of a nearby bush with their other hand; almost all of the individuals held the bush with their left hand and fished the oranges out of the water with their right. Apparently this task was tricky enough that in the monkey's brain one hemisphere took over the complicated tasks. The more difficult that an activity is, the more advantageous it is to assign it to a specialized part of the brain.

Once biologists realized that handedness could be a useful development, they started looking for it in other species, and when they asked the question in the right way (what is important and challenging to that species?), they discovered that the preference for a dominant side is present in most vertebrates. Right and left pawed-ness has been used to probe many dog traits, such as trainability and aggression. Other animals, from chickens and blue whales to honeybees and chichlid fish, have some things that they prefer to do on the right, while a minority of the species prefers to go lefty.

The question remains: If a specialized task could be better performed by dedicating one side of the brain to the job (and implementing it with the hand and fingers on the opposite side of the body), wouldn't it be better to switch hands for the left-hand dominant musician? We stopped trying to get lefties to write with their right hand, so why do we teach music as one-side-fits-all?

One obvious reason is that musical instruments are built to standard designs. If most pianists are more skilled with their right hand, it makes sense to build pianos with the treble notes to the right, where difficult finger gymnastics are required for melody and ornamentation. Trumpets are designed so that the fingers of the right hand can press the valves while the left hand simply holds up the instrument. And most particularly, stringed instruments are designed so that the left hand depresses the string while the right hand bows or strikes it to produce a tone. One might think that the left hand does the tricky parts, but picking, plucking, or bowing actually takes more finesse. Once the young violin student, righty or lefty, picks up their instrument for the first time, the teacher will say "This is how you hold it; chin on the chin rest, left hand under the fingerboard, and the bow in the your right hand." Any individuality is quickly eliminated. Teachers and instrument designers are the most important reason why everyone plays with the same hand, regardless of their preferred side. But not everyone does.

Charlie Chaplin was a passionate violinist, and not bad for an amateur; he even played it in several of his movies. But he learned to play left-handed, and had his violin altered, with the bass bar (a brace which supports the underside of the soundboard on the right, under the bass strings) and the sound-post (the round piece of spruce that props up the treble side under the bridge) reversed.

Paul McCartney was a lefty who switched from guitar to bass when the Beatles' original bassist left the band suddenly. Paul had to find a bass at a local music store, and re-strung the violin-shaped bass guitar so that it mirrored the usual instrument, with the lowest string at the top and the highest string at the bottom. And then there was Jimi Hendrix. Pictures of the rock virtuoso show him with a Fender Stratocaster, held left-handed and upside down. Jimi's father had tried to force him to play right-handed, superstitious that playing left-handed was the devil's work. The word "sinister" still carries some weight. But Hendrix was both ambidextrous and uncooperative. A search of the internet finds all sorts of claims about his handedness; some say that he could play a right-handed guitar strung normally, just the same as a righty. There are statements that he played a right-hand strung guitar, simply reversed to his other side, which would put the treble strings at the top and the bass strings at the bottom. Any guitar player will tell you that normal chord shapes are nearly impossible to play if the bass string is farthest away as the pick strums downwards across the strings. I spent an hour watching Hendrix videos on Youtube, trying to catch the closeups of his guitar strings and his fretting (in his case, right) hand. Indeed, this rock guitar hero played a Stratocaster built for right-handers, but he played left-handed, with the strings reversed so that the "normal" bass-to-treble relationship was just as it would be for a right-hander. The guitar still sounded different, however. The back pickup on an electric guitar is normally angled so that it is closer to the bridge under the treble strings, so that the higher overtones are emphasized, and angled farther away from the bass side of bridge to bring out the lower overtones of the bass strings. When Hendrix turned his guitar upside down to play lefty, the high overtones of the bass strings and the low overtones of the treble strings predominated. Whether this difference could be heard when playing at high volume through huge amplifiers is a different question.

You can get anything you want now days (if you are willing to pay), and left-handed models of guitars and other stringed instruments are available. Many rock players now play with their preferred hand, as well as bluegrass and country fiddlers. The common denominator is that these left-handed players are from self-taught traditions, where they figured it out for themselves. A left-handed fiddler would face serious obstacles if they decided to go classical and join and orchestra, however; the sight of 20 violin bows moving in unison would be in jeopardy if one maverick's stick was moving in the opposite direction. And the lefty's stand-mate might get poked in the eye.

There have been notable violinists who played their instruments reversed, the classical Jimi Hendrixes of their time. Finnish violinist Paavo Berglund learned to play on a violin his grandfather made him, and continued to play left-handed, even when he was a star with the Finnish Radio Symphony. The orchestra did have to rearrange its seating arrangement so that Berglund's bow didn't interfere with the other violinists.

Reinhard Goebel didn't have a choice: he learned to play violin right handed, but developed Musician's Dystonia in his left hand in 1990. Instead of giving up his career, Goebel reversed his violin and re-learned how to play from the other side. Later in his life, he switched back, sometimes even playing concerts as a "switch-hitter", some pieces from his left and some from his right.

Similarly, violinist Rudolf Kolishic learned to play right-handed and performed across Europe in famous string quartets, but injured his left middle finger and had to relearn to play from the left.

What if instrument makers and music teachers allowed left handers to switch to their dominant hand? In one study published in the interesting journal Laterality: Asymmetries of Brain, Behaviour, and Cognition (yes, there is a scientific journal for everything!), they built a keyboard that was the mirror image of a normal piano, with the pitches decreasing as the keys progressed from left to right. Volunteers who had never played the piano were given lessons on the reversed piano, and indeed the students who were left handed were able to learn more easily when the keyboard was adapted to them, rather than the other way around. This effect disappeared in left-handed students that had already taken lessons right-handed. The market for reversed keyboards is unlikely to take off (although it would be simple enough for an electric keyboard manufacturer to rewire their instruments for southpaws, if that was a thing).

Intuitively it seems strange that left-handers, although naturally inclined to play their scales and arpeggios from right to left, adapt to the "right" way of playing with such success. Legendary classical pianists Vladimir Horowitz, Arthur Rubenstein, and Glenn Gould never complained that their piano keys were arranged inconveniently for them.

Paul Wittgenstein was an heir of the richest family in Europe prior to World War I, but chose a career as a concert pianist instead of joining the family business. When he lost his right arm in WWI, instead of ending his music career, he commissioned concertos for the left hand alone and toured Europe with Ravel's "Concerto for the Left Hand", as well as works by Hindemith, Prokofiev, and Britten.

The one instrument that left-handers can call their own is the French Horn, but it isn't because some ambilevous (a sadly under-used term for left-handed) inventor intended it. The original horn had no valves, and the player held it with their right hand tucked inside the bell of the horn. Once valves were added, it fell to the left hand to operate them, since that hand wasn't doing anything useful at the time. Modern instrument makers have not made a switch to the right, and the valves remain on the left.

After this long trip thru the scientific literature, it appears that people who are left-handed are just as successful at music as right-handers, and even though they might do better if taught to play an instrument that was reversed from the standard right-handed version, they seem to do just fine, thank you. But why? Playing a musical instrument is one of the most demanding skills that human hands (and their corresponding cerebral hemispheres) are asked to do, and with most instruments the right and left hands have to work together to bring melody to life.

There it is: The hint we have been looking for.

The two cerebral hemispheres each have their own strengths and weaknesses, and have their own preferred appendages to carry out their instructions. But sometimes the right and left brains need to know what the other half is doing, and how to work together.

Enter the corpus callosum (Latin for "tough body"), a thick band of nerve tissue that crosses the divide between right and left cerebral hemispheres. This nerve bridge shares information between both parts of the brain; whether you are right- or left-handed, the information is shared with the brain and body's other half, a useful and necessary thing when the guitarist's right hand needs to pluck the string that the left hand is fretting, and do it within a hundredth of a second.

A robust corpus callosum seems like a valuable thing to have when coordinating two hands, so researchers imaged large numbers of brains of both musicians and non-musicians. As suspected, people who have studied music have a significantly larger corpus callosum than "regular people" (I started to write "normal people"). As a result of stronger communication between the right and left hemispheres, musicians score higher on tasks that require both sides of the brain to work together. A study of right-handed musicians showed less difference between their two hands than non-musicians, due to the increased development of the left hand. And left-handers typically showed an 11% larger corpus callosum than right handers. The left hand needs to know what the right hand is doing.

There is one group of people for whom the connection between the two halves of the brain is a curse, not a blessing. In epilepsy, an electrical storm moves across the cerebrum, causing life-threatening seizures. Although most seizures originate in one location in the brain, the electrical impulses sweep across the corpus callosum to ignite the neurons in the other half of the brain. When severe seizures cannot be controlled with medication, it is sometimes necessary to cut the neuronal bridge to keep seizures on one side from affecting the entire brain. When a callotomy is performed, most communication between the two sides of the brain is lost. Neurologists love this, because it offers an opportunity to study the right and left hemispheres independently. The result is that the right hand literally does not know what the left hand is doing. Imagine trying to to hammer a nail when the hand holding the hammer can't talk to the hand holding the nail? I searched the literature for medical cases of musicians who had their corpus callosum severed to treat epilepsy, to see if their musical ability evaporated; I could find no case reports and I suspect that callotomy patients have bigger problems than being unable to play the arpeggio introduction to "Stairway to Heaven". The loss of playing music may be the price they must pay to curb their seizures.

Jazz bassist Marc Johnson wrote and recorded a composition titled "Crossing the Corpus Callosum", an ode to the under-appreciated neural structure. The equal success of right- and left-handers in music suggests that communication between the hemispheres somehow levels the playing field for hand dominance. But what about those who simply listen to and appreciate music?

Since sounds that enter the right ear are processed by the left side of the brain, and vice versa, curious neuro-musicologists can play music to each ear separately to find how a preferred hand affects musical preference.

In one study, music was played to one ear only in 80 right-handers and 80 left-handers. Right-handers had more positive emotional reactions when the music was played to the right ear, and more negative reactions when it was played to the left ear. The preferences were reversed with left-handers. We rarely listen with only one ear, however.

Musical preferences also varied between righties and lefties. Right-handers generally showed a higher preference for more popular musical genres, while left-handers were more likely to dig less mainstream styles (reggae and bluegrass were the most significant). The study attributed this difference to the "increased cognitive flexibility" of left-handed people—the long-standing idea that left-handers are more creative and willing to consider different ideas. Right-handers showed less "inter-hemispheric interaction"—the traffic across the corpus callosum was decreased and the subjects were less "open-minded". This study involved music listeners, not musicians; since studying music increases the development of the corpus callosum, it is reasonable to suspect that all musicians, regardless of hand preference, would have a broader range of musical appreciation than the average listener, and interviews with accomplished musicians usually reveal startling tastes in tunes: one concert pianist listens to Pearl Jam in the car, Charlie Parker's favorite composer was Stravinsky, and Jon Bon Jovi dug Bach.

Playing a musical instrument is arguably the most difficult task that we ask of our fingers and thumbs. And yet we find ways to accommodate the demands that we make on our hands. Watching a performance of Gershwin's "Rhapsody In Blue", I can't guess whether the piano soloist writes with her right hand or her left; she can make magic with both hands together.

How High The Moon: Drugs and Music

When my son was in high school, I kept an ear out for the music he listened to. Nothing provokes parental concern as much as the phrase "sex, drugs, and rock'n'roll". When I heard him playing Huey Lewis and the News' 1983 hit, "I Want A New Drug", I started the obligatory lecture on the dangers of illicit substances. Halfway thru my "just say no" speech, he stopped me and made me listen to the lyrics:

"I want a new drug, one that won't make me sick,
One that won't make me crash my car,
Or make me feel three-feet thick.
I want a new drug, one that won't hurt my head
One that won't make my mouth too dry,
Or make my eyes too red".

Ok, point made; Huey Lewis' perfectly crafted pop-rock tune was a romantic song with a witty hook: He wants "One that makes me feel like I feel when I'm with you." Some people want more.

The history of music is full of musicians of all types who have explored a vast pharmacy of chemicals, which begs the question: Do musicians play better or worse when they are under the influence? And does their music sound better to us when we are under the same influence?
When this question came up, I first thought that surely someone had done a direct experiment: Record a pianist playing Bach's

"Goldberg Variations" or a saxophonist improvising over Coltrane's "Giant Steps" without taking any of the chemicals typically used for recreational purposes, and then have the same musician play the music after consuming a significant amount of alcohol, smoking a joint, sniffing cocaine, or partaking of an opioid medication. Then have a panel of musical judges listen to the recordings, blinded so that they are unaware of whether the musician was clean and sober or stoned out of their mind. Despite the easy availability of experimental subjects, I can't find that any controlled studies have been published. All we have is anecdote and opinion, which usually have more to do with the listener's inclinations than any objective analysis.

A corollary question is what effect drugs have on music listeners; do you really have to be stoned to fully appreciate Pink Floyd's "Comfortably Numb"? Or in reverse, is listening to psychedelic music a "gateway drug" that causes the sullen teenager to spending the summer smoking pot in their basement?

A search of the non-scientific parts of the internet (where millions of posts, blogs, and columns offer their own evidence for a position, for or against) finds an ocean of opinion. After a famous heavy metal band kicked their drug habits, one blogger states, their last 10 albums sucked, while another reviewer notes that The Lost Sessions, an album that saxophonist Stan Getz recorded near the end of his life after kicking his 40 year heroin habit, was his most inspired playing. I found it hard to draw any conclusion, either from the stories of artists so strung out that they had to be held up on stage as they sang and played, or from the scholarly dissertations about the drug metabolites in the wastewater at European music festivals. Has the world of music been helped or hampered by the ubiquitous use of drugs? Personal disclaimer break, not to condemn or recommend. I have played nearly every type of music over the past 55 years, including college rock bands in the late 60s (and yes, there were a LOT of drugs around), contemporary church choirs (the first choir director I played with resigned after it was discovered that she was smoking pot), jazz ensembles of older middle-aged guys (some admitted they used drugs when younger, but their wives made them stop), and community orchestras (I never asked any of the amateur violinists if they had experimented any further than a nice glass of cabernet). I have no direct experience to share, although I have administered a variety of brain-related drugs to my patients thousands of times. With humans, I only know what I have seen and heard.

From a physiologic and neurologic perspective, there are reasons why music might be better with chemical enhancement.
First, the various depressants (that's "downers", if you aren't a doctor) slow down the brain's functions and specifically inhibit the "executive functions" of the frontal cortex. That sounds like just what you don't want, but what we know about creativity is that too much thinking can be a bad thing. By turning off the "editor" in our head that constantly judges our performance, the parts of the brain that find novel connections between different ideas may allow the creative juices to flow. Pharmaceuticals that relax our inhibitions may also suppress the anxiety that plagues many performing musicians who suffer from stage fright, and

they feel they need a few drinks or a couple of Xanax before a gig (maybe propranolol, which doesn't affect brain function, would be preferable).

Stimulants (what the guy on the corner selling little bags of pills calls "uppers") go the other way. If you think faster, your fingers move quicker, and your energy level soars, which could be a good thing (if you can hold a steady tempo).

Hallucinogenics are anybody's guess. If you are seeing things that nobody has ever seen before, you may find that lost chord or paint exotic pictures with your lyrics that reveal other universes. But maybe the people in this universe don't really want to hear about it.

There is a downside to mixing chords and chemicals. Performing music is one of the most demanding skills that most people can do, and impairing movement and judgement could have unintended effects.

Oh, and you might die. Like Janis, Jimi, Prince, Charlie Parker, Bill Evans, country singer Luke Bell, and a long, sad list of others.

Still, the initial question stands un-answered: Do drugs help or hamper the concert pianist, the singer-songwriter, the jazz improvisor, the country fiddler?

Although my reading of druggie music studies was interesting, it didn't answer my original question. It did, however provide a lot of random findings, which I feel it is my duty to report:

1. When lyrics of hundreds of songs were analyzed for lyrical references to abused substances, the winner was not heavy

metal or rap, but country. Someone has been spending too much time in honkey-tonk bars.

2. Music festivals seem to be a prime venue for exploring the ubiquity of drug use. At a rock gathering in Croatia, analysis of wastewater during the festival week showed an overall 30-fold increase in MDMA and a doubling of cocaine, amphetamine, and alcohol metabolites. Intrepid researchers gathered urine samples from nine latrines at a Danish music festival (labelled as "pooled specimens", which conjures unpleasant images of the collection process) and found 77 different drugs in 44 samples. And in a popular concert gathering in the Colorado Rockies, investigators simply interviewed 17 attendees, who unanimously volunteered that they were using a variety of mind-altering substances. It seems that drug use is a given at outdoor music festivals. But what was it that drew the grad students who did these studies to sample latrines and sewage, and how did they convince their academic advisors that it was "behavioral science"?

3. Fan magazines relish the reports of famous performers that claim drugs as their muse. There is no lack of rumor, or fact. For Bob Dylan, amphetamines and cannabis were his thing in the 60s. Noel Gallagher and his band Oasis spent 3 years immersed in cocaine and alcohol while recording popular albums. Amy Winehouse, Lou Reed, Charlie Parker, Ray Charles—it is harder to find musicians without a drug history.

All of these studies are a lot of fun, but they don't answer my original question; they just confirm that music and drugs have a long history together. Then I realized that the question of whether drugs make for better or worse music was immaterial. The underlying issue is how we can judge the quality of the music we play and listen to.

This isn't to say that that there isn't good and bad music (and who hasn't heard both?), but unravelling the reasons that we love a piece of music is a vain attempt to look behind the magician's curtain.

Searching for artistic criteria for judging musical value, I ran across a blogger who suggests some possible ways to identify a good song:

1. It sends a chill down your spine or makes you feel emotional.

2. It gets your head bouncing or makes you want to dance.

3. It has a catchy melody that gets stuck in your head easily.

4. It makes you want to listen to it again and again

5. It receives positive feedback from other people or critics.

Other than number 5, all of these seem dependent on the listener's state of mind, and as far as reason #5, do you want to let other people tell you what to like?

Evaluating musical performances of conservatory students and aspiring orchestra players is more exacting, and often future careers are on the line. Judges need to have objective criteria to rank performers. One standardized checklist for rating piano students who accompany singers reads:

1. Uses correct posture and hand position
2. Introduces song
3. Cues singers to come in
4. Smiles and looks up when cueing
5. Plays correct chords
6. Plays chord changes at correct times
7. Sings along
8. Uses proper balance between the hands
9. Plays in steady tempo throughout
10. Continues in tempo if chords are missed

Consistency yes, but emotional communication and soulfulness, not so much. Numbered lists might tell us about music listeners' reactions or instrumental technique, but they don't provide a way to tell good music from bad. A more scientific approach is needed.

One study provides three musical ways of listening that hint at how music connects with its listeners:

1. Tone, or the hedonic quality of the musical timbre; When John Coltrane's saxophone plays the first note of his composition "Alabama", one can feel the indescribable feeling of sadness and beauty.

2. Dynamics, or changes in volume that demand our attention; Our brains are alert for any change that might mean that something significant is happening. The "heroic narrative" of Beethoven's later symphonies uses dynamics to paint the story of

a protagonist's struggles and victories in a way that can not be ignored.

3. Articulation, or the nuances of slurs, attack, and gliding between notes; expression provides the dimension of human vocal quality. When Stephane Grappelli plays his brand of hot jazz violin, his slides and staccato notes capture the risque enthusiasm of a 1940's French bistro.

These musical factors are great for program notes or music critics' descriptions, but what we really want to see is what the brain is doing that makes music connect with our emotions and our intellect. There must be some way to measure the effects of music on the brain.

The first reasonable approach would be to tape some electrodes onto a person's scalp and record the activity that is stimulated while listening to music. A study published in the Journal of Neural Engineering reported on just that, giving a detailed anatomical description of the effects. To quote, "cross-frequency coupling measurements were employed in a time-evolving manner to capture the evolving interactions between distinct brain-rhythms during music listening." I guess that is why they are engineers and not neurologists or musicians.

If they were neurologists, they would have reached for the functional Magnetic Resonance Imaging (fMRI) machine, which reveals oxygen uptake in the brain cells as they do their work. A group of neurologists focused on cells in the amygdala (the seat of emotions—now we are getting closer to the feelings of musical response) while the subject listened to either a jazz piano improvisation or composed piano music that imitated improvisation. Listeners were asked to judge whether the music was improvised or imitated, and they were correct the majority of the time (although they couldn't explain why). The fMRI readings showed greater response in the amygdala during the actual improvisations. The researchers suggested that this was because the improvs were more variable in timing and intensity—more unpredictable and more human.

Uncertainty is an integral part of jazz, but what about classical or popular music? A report in PloS One used fMRI to check in on brains while people listened to classical music, popular music, or a control. Both musical genres stimulated the right side of the brain, but researchers could identify the signatures of the genres by the anatomical location, classical music in the medial pre-frontal cortex and pop in the putamen. In other studies the putamen has been associated with "primary" reward (feel good neurotransmitters), while the mPFC is associated with cognition (intellectual interest), so the findings fit nicely with our musical stereotypes.

Our brain is made of individual neurons, and neurons are directed by the genes, so scientists who wanted to see what the genes were doing in those brain cells measured messenger RNA, the genetic instructions that carry messages from the chromosomes to working machinery of the cell. By sequencing the microRNAs of genes, they found that both playing music and listening up-regulated microRNAs (with catchy names like hsa-miR-132-39) that correlate with musical aptitude and other musical traits. Listening to music actually changes the genes in our brain (a modification termed epigenetics). This was true especially in individuals that were innately interested in music, but musical education did not show an epigenetic effect; maybe forcing children to take piano lessons does not create a love of music after all.

There are unexpected factors that influence our response to music. When musicians had to physically exert themselves during performance, they enjoyed the music more. Even non-players who exercised before listening found that the music sounded better. Sweating improves the tunes.

And it isn't just what we hear. In the journal Frontiers of Psychology a report described EEG evidence that "mirror neurons" in the brain responded vigorously while watching an opera performance. These special nerve cells produce patterns of activity that mimic the cells that would fire if the person was doing themselves what they observe another person doing. At first it was thought that audio input was triggering the brain to show a pattern of stimulation similar to the singer, but when they tested the music only, without the visual, the mirror neuron effect was lost.

This finding agrees with the conclusion of scientists from the Singapore Graduate Medical School in which the quality of a musical performance depended more on visual information than the sound itself. People were able to accurately predict the winner of a music competition by watching a video, even with the sound turned off. We may think that the strutting of the heavy metal guitarist or the dancing of Michael Jackson is just a big dose of showmanship, frosting on the musical cake, but it is part of the music and our brain likes it. Many people like listening to a string quartet more than they do a symphony orchestra; perhaps because the string quartet is more animated, leaning in to glance at each other frequently.

In a final fascinating look into how our brains judge the quality and enjoyability of music, scientists used "dual near-infrared spectroscopy" to investigate the relationship between a performer (in this case a violinist recorded on video) and 16 different female listeners. The degree to which the brain activity of the listeners matched the brain activity of the violinist ("left-temporal inter-brain coherence"), the more popular the performance was rated. They concluded that the audience perceived the performer's intentions thru a "temporally aligned network" and responded with positive emotions.

If quality can be defined as music that we appreciate, it appears that the factors that matter are its structure (the patterns of notes and rhythms that resonate with the brain), our personal history with the music (the meaning that comes from our experiences with the piece), and our appreciation of the technique required to bring the music to life (the velocity of the shredding rock guitarist or the unbelievably exact octave runs of the concert pianist).

Which brings us full-circle to the question of how drugs and music interact. Here are a few conjectures: On the musician's side, the brain under the influence might be freed to find musical roads less traveled, but the choice of which roads lead to good places might be suspect. When their music reflects their life experiences, it may lead us to dark places, where we find self-recognition (or warnings of danger). When it comes to technique, it is probable that pharmaceuticals impair the finely controlled movements needed to bring melodies and harmonies to life.

For the listener, the brain under the influence likely loses some awareness of the intricacies of tone, timbre, and the rising and falling of interwoven harmonies. If the audience shares the dark places of the addict's life (metaphorically or literally), those experiences may be emotionally potent. Fortunately, the listener who is high is not likely to notice the lapses in technique and expertise; (in my college rock band we would save our worst numbers for third set, when the audience had enough alcohol that they couldn't tell the difference).

Because perspectives are always different, we can't definitively answer whether drugs enhance or detract. Perhaps we should ask whether drugs help the person listening to enjoy the music they hear. There is certainly some music that would require an altered state to find something of aesthetic value. And, like choosing a dinner entree that pairs well with a vintage cabernet, those who choose to accompany their chemical highs with just the right sounds should choose wisely. Somewhere there must be an app that offers to "curate" your perfect choice.

Do drugs and alcohol make music better? For musicians and audiences alike, it's hard to tell. Maybe the answer is that music is so interwoven with human existence that it mirrors life itself, with its highs and lows.

Second That Emotion: emotions and music

The sky was black, the way it is at two o'clock in the morning in a small town in the farmlands of Eastern Washington. I had just driven back to the university from a gig at a bar across the border in Idaho (where the drinking age was 18, an attraction to the students at Washington State University in Pullman). The WSU parking lot was on the edge of the campus, a half-mile from my dorm; the deserted path to my dorm led thru the courtyard between the Gannon and Goldsworthy residence halls, featureless rectangular buildings where undergraduate students lived. Just as I was halfway between the two dark buildings, a booming sound seemed to come from nowhere, and everywhere: **Dum . Dum . Dum ....Da-dum.** The music was both familiar, and otherworldly: Richard Strauss' "Thus Spake Zarathustra" was familiar from its place in the opening of the movie 2001 A Space Odyssey, but the setting gave it a supernatural feeling; the waves of sound echoed from one building to the other over my head, making it impossible to tell where the sound was coming from. After a brief moment of anticipating an alien abduction, I realized that it had to be a prank, but I stood for a full minute in the darkness, feeling the full force of the music. A tingling sensation went down my back and the hair on my neck and arms stood on end. The music gave me the chills.

The next day I found out that the bassist from my band (who lived in Gannon Hall) had gotten back to campus twenty minutes ahead of me, put his huge bass amplifier in his open window, plugged his stereo into the amp's inputs, then waited until he saw me walking by to startle me with a 120 decibel wall of sound. The music had brought a quick succession of emotions: startle, relief, uplifting inspiration, humor. The chills.

There is a name for this sensation: "frisson", from the French, to shiver. Also termed "aesthetic chills", the feeling can be created by music or other forms of art and beauty, such as paintings, movies, or spectacular sunsets. This is just the sort of phenomenon that neuroscientists love to investigate, because frisson involves physiologic changes that can be measured. One study demonstrated how heart rate and blood pressure could be used to pinpoint exactly when a music listener became overcome with feeling. Another tracked the subject's eyes with tiny cameras, showing how emotional arousal could be defined by the dilation of the pupils. One study actually measured goosebumps on the people's arms with tiny little sensors. And some resorted to simply asking their subjects to describe what they felt at the climax of a dramatic symphony or during the high point of a tragic Spanish love ballad.

Armed with objective ways to measure frisson, the plan was to determine what exactly it is about Samuel Barber's "Adagio for Strings" or Prince's "Purple Rain" that gives us the shivers. Emotional, sadness-laden lyrics certainly help prime our emotions to be moved, but they don't raise chills on their own. Slower tempos and minor keys are often cited, and increasing volume definitely grabs us. However, researchers find that our physiologic frisson responses start just *before* the music peaks, when we suspect something is about to happen. Expectation, anticipation, and surprise seem to be the vital elements.

Imagine the original theme from Star Wars (if there is a master of goosebumps, it is certainly composer John Williams); the melody immediately jumps up a full fifth, a daring leap. As expected after a large ascending interval, the melody comes down three scale steps, one by one, and we feel like we are in safe territory. But then it jumps up a minor seventh, a bolder jump, almost an octave, but not quite. Then the second, third, fourth, and fifth notes of the theme are repeated, followed by the upward leap of a minor seventh (but now we hear it coming, anticipation). Finally, the second, third, and fourth notes of the theme are repeated (more expectation), but are followed by a step up and two steps down. Add the force of a full orchestra, with computer-generated stars whizzing by on the movie screen, and chills are almost guaranteed. Movie composers know about these things.
There are endless ways that musicians can create expectations and surprises that touch our emotions. The rock power ballad might start out with a long intro of fingerpicking on acoustic guitar before the drums come in, alerting our senses that something big is going to happen. then a repetitive verse and three-chord guitar pattern sets up a story for context. But the chorus is coming—you have heard the song a hundred times before, but it still works—the drums and guitars drop out for two bars as the vocalist takes the melody to a higher place; the rhythm comes back in (is that a Hammond B3 in the background?), volume comes at you like a tidal wave, and you feel the frisson. Have we been musically manipulated? Well yes, but art is about connecting with people's feelings, and those

emotions are hardwired into our brains and bodies by evolution for good reasons.

What good are emotions? (While it is tempting to work in a reference to Tina Turner's lyric here, I will resist it). From an evolutionary perspective, survival depends upon our two-stage response to anything sudden and unfamiliar, the same sequence that goes haywire in performance anxiety: First the amygdala, the emotional center deep in the brain, responds with a red alert: "Something unfamiliar just happened, probably dangerous." Signals are sent out to the heart, blood vessels, and internal organs to get ready for something, but we don't know what. Once the frontal cortex hears about this, our analytical brain says "OK, but let's think about this. There are no predators attacking or volcanos erupting, so it's all good". The body's arousal is already in full swing, but now we are reassured that there is no threat. Aroused and reassured. And it feels good, really good. We know, both from scientific studies and from daily experience, that when something good happens after something that could have been bad, we feel euphoric, whereas when something good happens after things have been going well, so what? Artists use this hard-wired response; the narrative of any great movie or book will have a crisis near the end, followed with a satisfying resolution. An emotional pop song may stack several expectation/anticipation/surprise/resolution cycles into four minutes. Even a good jazz improvisation will feature unexpected surprises that make us smile (once we realize that the musicians intended that dissonant chord or sudden stop), and when the band comes back to the familiar melody at the end, it feels like home. For those who don't have enough frisson in their lives, there is a website, ChillsDb.com, that provides speeches, music, and film clips that have been validated to induce chills in most people. There is also a Spotify playlist compiled by music experts to provide shivers and goosebumps; it includes samples as diverse as Beethoven, Hans Zimmer, Tracy Chapman, Led Zeppelin, and Death Cab for Cutie, all with the potential to send shivers down our spines.

To dive into the frisson experience, (and see if it matched the theoretical criteria above), I chose to check out Phil Collins' "Against All Odds". The verse is expected: piano playing one chord on each beat over a smooth synth pad, no big surprises in the melody, story of heartbreak in the lyrics, each verse ends in the line "You are the only one who ever knew me at all…", a symmetry that anchors the verses. Then, volume comes up and the piano plays with more urgency, "But take a look at me now." The lyrics pause for a bar, and "Cause there's just an empty place"—this time the chords are a surprise, moving from a C major up to D major, an unexpected key change, before coming back into the key of C with an A minor, F, and D minor chords. Beautifully constructed ballad, with just enough unexpected twists of melody, dynamics, lyrics, and harmony to capture the feeling of heartache and goosebumps. I didn't actually experience any frisson while listening to Phil Collins sing, but that's what you get for trying to be analytical when you should just be following the feeling.

It seems to be a universal that sad songs are more likely to give us emotional chills than happy ones, but paradoxically they make us feel more comforted and happy than more upbeat songs. Distress followed by reassurance.
The emotions triggered by music have complex components. The amygdala/frontal cortex polarity plays a part, but all sorts of other cognitive processes are also at work: Memories of having heard the music in other contexts, instinctive reactions to the harmonic intervals, and the timbre of the instruments (the more closely they coincide with the human voice, the more they affect us). Cello is the quintessential voice-like instrument, and when his producer suggested to Dan Fogelberg that they add a cello track to one of his songs, Dan responded that "This song is already sad; if we add cello, somebody might die from it.".
Trying to trace the pathways that connect music and emotions is futile (unless you are a neuroanatomist with an MRI machine sitting idle). Nearly every center in the nervous system seems to be involved in some way. Even the chemical substances that carry the brain's feelings thruout the body seem to laugh and weep. Dopamine, which acts as a neurotransmitter in the brain, also functions as a neurohormone when it is released by the hypothalamus, influencing learning, memory, mood, and movement. It is most often referred to as the reward hormone, surging during sex, drugs, gambling, and yes, music. It is the feeling of expectation and reward.
The opioid system also responds to melody and rhythm. Daniel Levitin and his associates tested how endorphins play a part in the emotions elicited by music by giving naltrexone (an opiate

blocker, like naloxone) while subjects listened to music that should have moved them, and found that the feelings were absent. This anhedonia (lack of emotion) was both positive and negative; happy music brought no joy, but sad music brought no sadness. As I read this study, I pictured emotional concert-goers carrying an anti-narcotic Naloxone injector, just in case they got too carried away.

Musical emotion also influences how we perceive our world. In another study coffee drinkers were asked to rate two different coffees, while listening to silence, a "sweet" musical fragment, or a "bitter" musical piece. Bitter music made the coffee seem bitter, while the sweet melodies were associated with joy, surprise, and a little sadness; they also made the coffee taste less bitter. (Are you listening, Starbucks?)

Unfortunately for our emotions, we rarely really listen to the music around us; it has become merely the ubiquitous background of modern life. While we used to use "elevator music" as a derogative term for bland sounds, we are now exposed to it in our workplace, on television (and not just in the ads), at the airport, and even during the interminable on-hold experience while we are waiting for the receptionist at our doctor's office to return to the phone. These are not emotional experiences (well, maybe we have intense emotions while we are sitting on hold, but that doesn't count), and the use of music devalues its effectiveness.

The power of emotion as a therapeutic tool has not been lost on the healing professions. Guided Imagery with Music (GEM) is one form of music/psych therapy that uses emotional music along with verbal prompts from a therapist to address issues from depression to post-traumatic stress. The use of emotional music seems to allow talk therapy access to parts of the the psyche that are otherwise unreachable. Strong Experiences with Music (SEM) uses more music, less talk: The patient is asked to listen to selected music without distraction. Initially the music is matched to the mood of the person,(let's say Billy Strayhorn's "Lush Life", Johnny Cash's "Hurt", or even Beethoven's "Moonlight Sonata", to catch the depressed mood. Then the music gradually progresses to neutral, then hopeful, and finally uplifting (Pharrell Williams' "Happy", Bob Marley's "Three Little Birds", or "Spring" from Vivaldi's "Four Seasons"). The mood of the listener is pulled along by the changing emotions in the music; sometimes it is uncomfortable, and painful thoughts and tears are to be expected, but eventually the clouds are lifted. The Indigo Project is a mental health organization in Sydney, Australia, which includes a psychology practice and a creative workshop academy. The Project has developed an immersive music mindfulness program called Listen Up to harness the emotional power of intentional listening in a protected setting. Participants in Listen Up are invited to rest on mats or cushions in the studio as the lights are dimmed. Carefully chosen music starts as a facilitator asks the listeners to listen deeply to the sounds, which include selections of ambient music, movie soundtracks, and experimental compositions. People are encouraged to listen with attention, even while they maintain awareness of their bodies, thoughts, and emotions. After the fifty-minute session the lights come up gradually and the participants are invited to pause, reflect, and chat with their neighbors.

The Listen Up program was studied in an effort to explore the effects of attentive listening. As with SEM, some found it unsettling, as it called up unpleasant memories and negative emotions. Others simply found it difficult to maintain focused attention. Most found that the experience brought out intense and complex feelings. Afterwards, however, all of the listeners were positively affected. Many of the comments afterwards used words like "cathartic" and "epiphany", and when moods were rated on a standard scale they had improved dramatically. Many said they were transformed, and reported transcendental, life-changing experiences from a single session.

Feeling better is good, even if it is only during the afterglow of an hour of listening, but the Listen Up sessions seemed to have lasting effects. Even weeks later the participants continued to feel more calm, more grounded, and more confident about the direction of their lives. This effect seems similar to the newly-popular use of psychotropic medications like ketamine or psylocybin, where even a single use will sometimes lastingly change a person's emotional outlook, lifting the burden of depression, loneliness, and trauma. It seems that we already have the power, to change if only we could pay full attention and listen to the music.

Now, if I could only turn off the on-hold music from my doctor's office.

Over the Rainbow: music and meditation

An empty, weather-beaten 600,000 gallon water tank in Rangely, Colorado (pop. 2296), on the western slope of the Rockies, seems an unlikely place to hear Beethoven. The Tank is seven stories tall, and its vacant interior has been turned into a unique (and somewhat eerie) performance space. What makes it special is the sustain and reverberation; a tone played in the center of The Tank continues spinning around the circular space for 43 seconds before fading away, which creates an overlay of one note on another that cannot be achieved in a recording studio.

Music played in The Tank naturally takes on a suspended feel, but the effect is exponentially more timeless when a string quartet plays the "Adagio" from Beethoven's "String Quartet in C# Minor, Op. 131"—but slowed down to one-sixth of the normal tempo.

The Slow Beethoven project was driven by Jeffrey Ziegler, former cellist with the Kronos String Quartet. When the Tank Center for the Sonic Arts approached him, he agreed to put together a string quartet to explore the sonic space. As with all the best laid plans, the covid pandemic interrupted the process, forcing the musicians to perform the music at a space in Brooklyn and send the audio feed in real time to The Tank, where high-tech sound equipment projected the music into the water tank as if the musicians were there. The sounds were captured with state-of-the-art recording equipment, and the result was Slow Beethoven, Music from The Tank.

Listening to Slow Beethoven, with no distractions, is a powerful and unique experience. In the words of Zeigler, "the sound activates an array of harmonics that makes each passing moment feel like the sound itself is alive." Because of the long delay of each note, the melody of Beethoven's composition disappears, leaving only texture and timbre. For 44 minutes, it feels unmistakably like meditation.

Music and meditation seem to be a natural and complementary pairing. Both music and meditation offer fascinating material for research. For one thing, many of the subjects are tranquil and compassionate people full of mindfulness, which is more fun than studying the brains of psychopaths and abusers. Study subjects are also mostly "normal" people, rather than patients who have had strokes, brain tumors, or severe blows to the head. Much of the interest in meditation (and in music) centers around the electrical activity of the brain. So-called "brainwaves", the spikes of electrical activity in the brain, are one way to visualize how the brain functions during different activities.

In the 1920s, German neurologist Hans Berger discovered that he could detect weak electrical signals from the brain by gluing electrodes onto a person's scalp and recording the impulses. At first these signals were used to study the electrical storms of epileptic seizures, but he found that even the brains of normal people produced rhythmic jagged lines on his tracing equipment. Brainwaves happen when groups of neurons fire in a coordinated manner, and the waves are classified by the frequency of these oscillations: Delta (1-4Hz), theta (4-8 Hz), alpha (11-12 Hz), beta (above 13Hz), and gamma (30-40 Hz). If your brain is dominated by delta waves, you are probably asleep, but theta waves also occur during sleep (especially when dreaming) and during focused relaxation. When alpha waves predominate, your mind is clear and relaxed; when the beta waves take over, you are alert and active, the normal waking state. Rapid gamma waves are said to indicate higher levels of consciousness. As with most scientific discoveries that make their way into the popular consciousness, brainwave differences were quickly used to explain everything and soon gadgets appeared on the internet that claimed to measure and tweak the neural pulses that make our brain work. Despite the hype, there is a lot we can learn from listening to the electrical activity of our brains.

The "default mode" of predominantly beta waves helps us get a lot done, but also creates the "monkey mind" that makes us feel overwhelmed in everyday life. The faster gamma waves can fuel even more frantic brain activity It seems that if we want to relax, de-stress, and focus attention without distraction, we want to slow the brain into the theta and alpha ranges.

Reading the brain's frequencies offers a way to understand what happens in the brain during mindfulness meditation. When experienced meditators were studied initially, the focus was to discover whether the mysterious practices of Tibetan monks and Christian mystics actually altered the way the mind worked. Typical of hundreds of studies using different types of meditation and yoga practices, one study of Vipassana yoga found a marked increase in theta and alpha waves during meditation. Other studies involving Transcendental and Zen practices found that the slower oscillations of theta and alpha increased in strength as beta waves decreased. In some of the more intense traditions, the faster gamma waves were also increased, and were correlated with more intense spiritual experiences.

Can music alter brainwaves in the same way that meditation does? There have been lots of studies of music and mindfulness, but they usually focus on easy-to-measure physical responses like heart rate, blood pressure, or galvanic skin response (a measure of stress, used in lie detectors). Indeed, under the right conditions music consistently reduces anxiety and physical measurements of stress. In one study, live improvisatory viola music lowered the brain wave frequency from 12-13Hz (beta) to 8-9Hz (alpha). In other tests music increased alpha activity in posterior parts of the brain, and imagining music had an even greater effect than listening to the real thing. It seems that the right music can offer many of the same benefits as mindfulness yoga and meditation.

As always, it depends on the type of music. In the research setting, music without words is always used, to minimize the distraction of lyrics; when language gets involved everything gets murky. Classical music is the the usual choice, but "ambient" music, drones of spacey synthesizer sounds, or the moaning of a wooden flute are often used to produce calming effects. Not surprisingly, slower music was found to be more relaxing than faster music. The experiment mentioned above used viola because it closely matches the human vocal range, although in my opinion it would be hard to do better than Bach's solo cello suites if you want to zone out. The cello suites may have a lot of notes, but they flow so naturally that the brain is not asked to analyze, just ride the river of music as it goes by.

An ambient music group from England used what we know about the effects of music to record a composition titled "Weightless", billed as "The World's Most Relaxing Song." The tempo starts at 60 beats per minute, the average resting human heart rate, and gradually slows over 5 minutes to less than 50 bpm; through a process called "entrainment", the tempo pulls the brainwaves to a slower frequency. In this piece, nature sounds (birds, waves) are mixed with guitar and synthesizers to produce a dreamscape so powerful that they recommend against listening to "Weightless" while driving.

In my own listening comparison, the Slow Beethoven recorded at the Tank Center induced a more meditative state of mind than "Weightless". This may be because in the Beethoven the melody is there, but it isn't. In order to perceive melody in a pattern of notes, the tones need to be held in short-term memory; however the limit of short-term memory is roughly 7 seconds. If a sequence of musical tones stretches beyond this limit without a pattern emerging, our brain hears the notes, but can't make them into a melody. The same trick is often used in movie and television music when the music is intended to provide mood and texture while remaining inconspicuous and not drawing attention to itself. When I listen to Beethoven's "Adagio" played at a normal tempo, I count about 8 notes within my 7 second short-term memory, and a slow but pleasant melody floats by. In the Slow Beethoven recording, which was played and recorded at one-sixth of the usual tempo, the notes barely change once during the span of my short-term memory; each note also shares the sonic space with reverberations of all the notes that were played by the violins, viola, and cello during the previous 43 seconds. The music is beautiful in its own way, and grounds the listener mindfully in the current moment, no thinking of past or future, as the music seems to exist outside of time.

Mindfulness meditation and music share some of the same effects within the brain, and many practitioners of Zen, yoga, and other traditions use music to enhance their experience. Others find that even calming music distracts their minds and they prefer to meditate in silence.

Music is by its very nature ephemeral; each note is heard for a moment, and then it is gone. This should provide an antidote to the jumble of competing stimuli that make up our everyday lives. But it doesn't. The reality is that we rarely "listen" to music; the car radio cranks out pop tunes while we are driving the kids to school, Starbucks is playing an esoteric mix of musical genres while we are trying to order our grande soy milk latte (extra hot), and the suspenseful music on the TV murder mystery only causes us to yell at the TV, "Don't open that door! Can't you hear the scary music?!" None of the studies about music and alpha brainwaves used real-life scenarios, and I am guessing that listening to "Weightless" while standing at the counter of the Department of Motor Vehicles won't lower your blood pressure or put you in the moment; you are already wishing that this moment would hurry up and move on to a better moment.

There are many facets of music that improve our mood, and these are vital parts of the music therapist's toolbox. Familiar music can stimulate memory in the elderly, relaxing sounds can decrease the perception of pain, and a patient's favorite tunes can reduce anxiety. Presumably these effects are actually changing the electrical and chemical activity that is the basis of our minds. Who better to embrace the mind-bending interrelationship of music and meditation than musicians? Music performance is challenging and often terrifying, even to those who have made it their life. Many musicians use familiar routines like scales and arpeggios that are nearly automatic to relax before a concert, or to put themselves in a mental space for a demanding practice session; the music becomes the meditation. Other meditation practices are often used for performance anxiety to clear away distractions and create a space of self-compassion to remind the players that they are performing in the service of their art, not of their egos.

Relaxation also releases musical creativity. But too much restfulness might not be helpful. Leonard Bernstein tells that he often will lie down and relax when he is trying to find an idea for one of his compositions; unfortunately, he reports that he usually just falls asleep.

Both meditation and music have powerful effects on the mind, but they aren't magic. Calm attention to a single stimulus, whether it is attending to to the ebb and flow of one's breathing or to the patterns that sound waves create in the auditory cortex, changes the electrical pulses in the brain. I wonder if the same effect could be measured in the camper who sits gazing into a campfire in the evening darkness, watching the orange flames lick at the burning wood and the glowing of the coals underneath. It seems like we should seek those moments, whether they are gazing into the fire, practicing mindfulness yoga, or sitting in a quiet room listening to a classical guitarist play Erik Satie's "Gymnopedie".

Sweet Dreams Are Made Of This: sleep and music

Sometimes science comes late to the party, elaborate studies only confirming what our species has been doing for millennia. In a village in a remote area of New Guinea, a baby's cry is heard, followed by the soft sing-song of a mother's voice. Within minutes peace is restored and the infant is folded into a soft blanket of sleep. The lullaby has worked its magic.

Lullabies in all cultures tend toward slow tempos, low volumes, and breathy words. Even songs in a foreign language have the desired effect (isn't every language foreign to a newborn?). It seems universal that mothers know how to sing their babies to sleep.

Enter the scientists, who wire up babies in the hospital NICU to monitor heart rate, oxygen saturation, sleeping time, and cortisol levels while playing recordings of spoken voice, rhythmic sounds, recorded lullabies, or a mother singing live to her child. These studies have revealed that the lullaby decreases the baby's heart rate and cortisol level and improves oxygen levels and sleeping time.

Not every baby follows the script and some remain unconsolable (or maybe some mothers are no Taylor Swift) but many of the studies expose infants to lullabies for only two minutes at a time, three times a week; every mother knows that it takes both soft music and a large amount of patience. If nothing else, singing calms the mom and distracts her from the crying child in her lap until sleep takes hold.

Since the child is the eventual adult, the next obvious question is whether music can provide the same drowsy bliss in adults; in one survey, 46% of respondents stated that they sometimes used music to drift off, but direct experience is never enough proof for the music researcher. Let's put people in the sleep lab, wire them up to record brain waves, monitor their hearts to record rate and pressure, and check their blood cortisol levels—then ask them to listen to music as they try to drift off to sleep. Perhaps this isn't a real-world setting, but at least it generates some data about the effect that music has on slumber.

In general, listening to music while falling asleep or staying asleep improves sleep quality. Preschool children take better naps while listening to classical guitar recordings. Grade school children sleep better when listening to music. Nursing school students show decreased sympathetic nervous system activation (fight or flight), lower heart rate and blood pressure, less anxiety, and improvement of depression symptoms when sleeping to music. Pregnant women rest better, and patients hospitalized for angiography drift off more quickly. And elderly individuals, who often lose the ability to get a good night's sleep, find relief in melody and harmony.

Insomnia can be its own special torment, and patients and their caregivers are willing to try anything to get a night of sleep. Melatonin supplements have been tried by millions of people, and sleep medications (both prescription and over-the-counter) are widely used, despite their serious side effects (including significant increases in cancer). Home remedies, like exercise, a hot bath, and a warm glass of milk at bedtime are no match for the truly sleepless night. Naturally, music interventions hold inexpensive and accessible promise. But, what kind of music? Internet advice-givers repeat the obvious: slow tempos, low volume, acoustic instruments without lyrics. Songs that make you relaxed and happy are supposed to help you fall asleep faster and stay asleep longer. But that isn't always what sleep studies have shown; happy or sad music worked better than no music at all, and the results were best when people chose their own music, regardless of genre. Even when the self-chosen tunes were exciting, rhythmic, and featured screeching guitars, sleep came more easily than lying in a silent room.

If we want to find out what people really listen to as they doze off, we have an enormous data bank to mine: Spotify.

The music streaming service has over 1000 playlists that listeners have compiled to help them sleep, which allows data sleuths to determine which types of music are preferred at bedtime. As expected, familiar classics like Brahm's Lullaby, Cannon in D, and Claire de Lune show up frequently on the playlists world-wide, but half of the lists have more energetic tracks in a variety of genres. K-Pop, boy bands, electro-pop, and lo-fi beats were well represented. Similar surveys in the UK and Australia included 20-30% classical, 20% rock and pop, 10-20% acoustic and folk, and a variety of ambient, soundtrack, and jazz. The song that appeared most often on Spotify sleep playlists was "Dynamite", by K-pop band BTS. The surveyors were surprised and a little puzzled by the wide variety of music that helped people sleep.

It seems to be an accepted fact that music and sleep make good bedfellows, but we can't really understand how to use music unless we know why, and there have been many hard-to-prove suggestions. One scholarly paper examined what they termed RPRs (Researcher Proposed Reasons; in other words, wild guesses that the sleep scientists pulled out of thin air). The top six RPRs were 1. Relaxation (and who doesn't need to relax when they are trying to sleep?). 2. Distraction (watching reality TV in bed would also distract, but has its own drawbacks). 3. Entrainment of biological rhythms (the rhythm is gonna get ya). 4. Masking of background noise (like your roommates watching reality TV in the next room). 5. Enjoyment of emotionally pleasant music (you feel better because you feel better?). and 6. Cultural expectation (you will sleep because we all told you that you would).

Even these very scientific-sounding RPRs must have some physiologic basis; our brain listens to our body (well, at least some of the time), and the physical messages that the incoming nerves bring to the brain have more influence on the nervous system HQ than do our conscious thoughts. This brings us back to those brain electrodes and blood tests. When music is playing, our stress-induced cortisol decreases, our relaxing and rewarding dopamine increases, the parasympathetic nervous system kicks in to decrease our heart rate, the Heart Rate Variability increases (which doesn't sound like a good thing, but it is), the blood pressure drops , and the Skin Galvanic Response lowers (in other words, don't sweat anything). With all of those chill incoming messages, the conscious brain can clock out and grab some much-needed sleep.

It is an axiom in medicine that any treatment that actually does something can also have side effects (when we read on the internet that some alternative medicine supplement will cure everything with absolutely no side effects, it is time to turn off your laptop—it is probably keeping you awake anyway). A cautionary tale about music side effects was uncovered in a study conducted by Dr. Michael Scullin at Baylor University's Sleep Neuroscience and Cognition Laboratory. Dr. Scullin noticed that he would sometimes get "ear-worms", in which a familiar song would get stuck on repeat in his brain when he was trying to sleep. He wondered if this was a common phenomenon, so he intentionally induced ear-worms in volunteers before sending them to sleep in the lab (doesn't Baylor have a research ethics department? Who would do that to people?). The researchers played Taylor Swift's "Shake It Off", Carly Rae Jepsen's "Call Me Maybe", and Journey's "Don't Stop Believin'" while the subjects were trying to sleep. They also played instrumental versions of the same tunes. The sleepers reported experiencing ear-worms as they tried to doze off and they frequently noticed them when they woke during the night. When they awoke in the morning, the ear-worms often returned. Many of the volunteers had experienced ear-worm interference when they were sleeping at home, and for those people the laboratory measurements confirmed that they took longer to get to sleep, wakened more often, and spent more time in light sleep than normal when bombarded with catchy melodic tunes before sleep. Researchers found increased activity over the areas of the brain where memory consolidation takes place during sleep, and concluded that the brain was trying to process the music that wouldn't leave the brain. Curiously, the instrumental versions of the songs created more sleep disturbance, as if the memory centers of the brain were trying to complete the familiar melodies by searching for the words. On the EEG tracings, those sleepers who experienced ear-worms also showed more slow oscillation brainwaves, indicating reactivation of memory centers during sleep. This reactivation revealed that memories were being transformed into long-term traces. Clearly, our brains continue to process music, causing it to linger in our minds.

If our brains are still active during sleep, does that mean we can learn from what we hear (or imagine) in our sleep? This is an attractive idea; who wouldn't want to learn a foreign language or study for a history test while grabbing some zz's? The concept of "sleep learning" enjoyed popularity (and commercial exploitation) in the 1930's, based on studies by German psychologist Rosa Heine. She found that material studied just before bedtime resulted in better recall the following day, so a variety of machines were sold to expose sleeping students to study material. It seemed that there might be a shortcut to cramming knowledge into our brains. But when the electro-encephalograph was invented in the 1950s, researchers found that the sounds that students heard were actually waking them up briefly, and later recall was related to their alert state, which didn't do much for their sleep.

What goes around, and all that: More modern studies have shown that there is something to sleep learning, and it has to do with music. When students studied something boring (like macroeconomics, for instance) while listening to a specific piece of classical music (like Chopin's Nocturne in C# Minor, for instance), the economics memories could be reactivated and transferred to long-term memory if the same music was played after the student had reached deep sleep. Called "targeted memory re-activation", this new twist on sleep learning shows that memories can be replayed during sleep and transferred from the hippocampus to other parts of the brain for later recall. The idea sounds interesting, but whether learning economics is worth listening to Mozart, getting wired up, and having someone play the music after you get solidly asleep is questionable. The nature of the information being studied also matters; implicit memories, the skills we learn that we don't have to think about, seem to be open to sleep learning, but explicit memories, like what economic theory John Maynard Keynes is known for, are less susceptible.

In this book we are not concerned with macro-economics, so what we really want to know is if we could learn to play Chopin's Nocturne by listening to it while we sleep. Well, sort of. To test this idea, people with little or no musical background were taught to play two simple melodies in response to abstract symbols in a computer game (think of the game Guitar Hero). Then as the subjects napped, one of the two melodies was played. After waking, the person played the game again with each melody to see if they had improved. Even though they did not remember hearing music while they slept, they scored better on the tune that was played during slumber. This won't get you to Carnegie Hall, but there is evidence that mastering a piece of music that you are working on could be enhanced by hearing it in your sleep, encouraging your brain to consolidate the related memories.

And some people dream music. Musicians experience music in their dreams more often than others, and the more time spent during the day playing music, the more that the sounds show up in dreams. But on the whole, dream surveys show that only 8% of people's dreams have musical elements, although a different study revealed music in 20% of dreams, even for people who had never played an instrument. As people age, however, the amount of music soundtrack in their dreams declines, perhaps because of less time spent listening to music.

Most people report that when music occurs in a dream, so do more positive emotions. Psychologists asked a group of grad students about their music preferences, and then monitored the types of dreams that they reported. Heavy metal fans had more "dissociative avoidance" dreams, as well as more frequently dreaming that they were dreaming or unconscious, while those who preferred rap had more sexual dreams. Jazz fans had more recurrently pleasant dreams, while classical aficionados dreamed of flying, but felt less contented in the dreaming adventures. These correlations may be more related to the peoples' personalities and life stages than any influence of their preferred music, since the described dreams were just preferences, not the results of listening to the music as they slept.

Hearing music late at night might influence the dreams that follow, but it is more interesting to consider when the reverse occurs. In one report, half of the music that appeared in dreams was not familiar to the dreamer; perhaps the melodies were generic enough that they lacked an identity, but in many cases, especially among musicians, original music was being generated while the musician slept. If the person could capture what they heard in dreams after they awoke, new music could be created. Among classical composers, dreams were a common inspiration for their pieces: Debussy's "Reverie", Faure's "Apres un Reve", and Liszt's "Liebestraum" invoke dreamlike imagery (although we don't know if these pieces came to the composers in their dreams). But because we know everything about the lives of contemporary popular musicians, there are dozens of stories about songs that came unbidden during sleep and were captured, recorded, and immortalized into rock classics. Sometimes a melody appeared unbidden in a dream, like Paul McCartney's "Yesterday", to which lyrics were added after waking. Researchers often find that music takes precedence over words in sleep, suggesting that music is more basic to our human brain (and lives more in the right hemisphere, releasing creativity while the language-based left hemisphere is sleeping). Other times a lyrical hook comes with the tune; Keith Richards awoke with a riff and the words "I can't get no satisfaction" in his brain and a classic was born. Many of Billy Joel's hit songs came to him in dreams, often fully formed and accompanied by synesthetic colors. Jimi Hendrix' "Purple Haze", Paul McCartney's "Let It Be", and Todd Rundgren's "Bang The Drum

All Day" came from dreams, testifying to the creative unconscious activity that occurs while we are sleeping. Sleep, dreams, and music are a magic subconscious playground. Part of my reason for writing about music and the brain is a search for insights about music and my own brain. I don't remember being crooned lullabies as a baby, but I have spent a lot of waking time since then playing, writing, and thinking about music. I have never had a dream accompanied by music, however. It would be nice to have beautiful music appear during sleep, waiting to be written out and played, but this experience has eluded me. But I can dream, can't I?

Read Between The Lines: written music

When New Orleans trumpeter Louis Armstrong was asked if he could read music, his memorable response was "Sure, I can read music, but not enough to hurt my playing." The relationship between reading music and playing music is laced with tension; "trained" musicians often dismiss those who play "by ear" as being second class musicians, while popularity (and musical fortune) often come to the guitar player, singer, or even composer who is unable to decipher the clusters of black ovals that balance on five parallel lines.

The list of musicians who are unable to read written notation is long and recognizable: Guitarists B B King and Jimi Hendrix relied on their ears to find the perfect notes. Songwriters Paul McCartney, Bob Dylan, and Prince used the chords they had learned by their shapes to create the most memorable songs of the 20th century. Irving Berlin, who wrote many of the tunes in the Great American Songbook, could not read music, and he could only play the piano in one key (he had a hidden lever that changed the positions of the piano hammers to shift keys, in the same way that a guitarist uses a capo to change their three chords up or down in pitch). Composers like BJ Liederman (who writes most of the theme music heard on National Public Radio) and film composers Vangelis (Chariots of Fire), Danny Elfman (Spiderman, Good Will Hunting), and Hans Zimmer (Pirates of the Caribbean, Dune) recorded music from their keyboards and had them transcribed for the musicians who performed on the soundtracks because they never mastered the skill of reading music..

Even in university music programs there has been an increase in students auditioning as music majors who are unable to read music; they can play prepared music well enough to be aspiring professionals, but can't decipher the spots on the page. Similarly, directors of classical music camps and ensembles find themselves teaching unfamiliar music with the "oral/aural" method: "It goes like this—da, da dut da", until the students learn their parts by rote.

Every summer, hidden in the pines and oaks of the quaint mountain town of Julian, California, the Julian Family Fiddle Camp welcomes acoustic musicians of all ages, instruments and abilities. Players spend the week learning and playing bluegrass, folk, and Celtic tunes in the brisk open air. Visitors will notice fiddles, guitars, flutes, and mandolins, but they will not find one piece of written music. Participants are encouraged to learn songs the way that they learned to speak as toddlers: Listen, imitate the sounds they hear others making, and join in as soon as they feel able. Most of the musicians are there for the fun and camaraderie of playing traditional music with others, but there are also a number of classical string players who come to reconnect with music that doesn't come from a page.

Throughout most of most of musical history, and in musical cultures across the world, music has been passed from one musician to another by one person humming a tune while the other person tries to match the notes on their flute or fiddle.

Most folk music, Celtic tunes, and rock'n'roll songs are learned by listening.

Efforts to transfer musical knowledge by writing it down is not a new thing, however. The oldest written music is found in cuneiform writing on clay tablets from Babylonia, dated 1400 BCE. The marks visually represented which strings to play on the lyre. Then every age and culture added its own improvement; Greek song texts came with letter symbols above each word to indicate what note the singer was to use. Musicians in the Byzantine Empire represented the melodies of Orthodox chants by indicating how many steps up or down the song moved within a pre-established mode or scale. By the 9th century CE, monks in Europe were drawing highly decorated books of Gregorian chant with square notes that moved up or down, but without set pitches. Credit for "inventing" our standard musical notation in the 11th century goes to Guido d'Arezzo, who placed notes on staff lines, and named them for the first syllables in each line of a Latin hymn: Ut (later Do, for Dominus), mi, fa, sol, la, and si. (Admit it, you are picturing Julie Andrews teaching the Von Trapp children to sing).
Reading music from a page (or a clay tablet) has been around for more than three millenia, but there is still a wall separating the musically literate from the musically rich and famous. Superstar singers Elvis Presley, Michael Jackson, and Taylor Swift have had legendary careers without looking at a piece of sheet music. Eric Clapton may have his own corner of the Rock'n'Roll Hall of Fame, but he can't read the music that he

plays. One can't help but wonder if learning to interpret written music notation is worth the effort.

Admittedly, reading music is hard. In some ways it is similar to learning to read words. Certain brain areas overlap between the two skills and accomplished music readers share significantly more brain areas for reading words and music than novices. Researchers have studied the differences and similarities, partly to understand how children might be taught to read words more easily; if we all have neural strategies for decoding notes on a staff, maybe some of those could be linked to understanding words on a page. In both, general intelligence helps, but in literate musicians, "domain-specific" skills are more important; recognizing complete patterns that the musician understands ("It starts on the third note of the major scale and goes up by whole steps, ending on a major triad") becomes more important than simply decoding the individual notes.

Dyslexics take longer to learn musical notation, but tend to devise strategies to compensate, including a better musical ear. Similarly, when dyslexics struggle to read words they also use tricks and work-arounds to overcome (and sometimes hide) their difficulties.

Maybe it shouldn't be that hard. When asked, 90% of 6-year-olds believed that they could write music; I observed that my own granddaughters, when they were 2 or 3 years of age, would cheerfully make marks on a piece of music manuscript paper to show the notes that they had picked out on the piano. I had to take their word for it when they played what they had written. However, only 15% of the 6-year-olds who believed that they could write music also thought that they could read it. Writing is easy, reading is hard.

Reading music presents more obstacles than deciphering words. The first thing the music reader needs to do is check the key signature at the left edge of the music to remember what notes may be flats or sharps; The reader of words doesn't have to check the beginning of a paragraph to see what language it is in. Unless the reader of words is reciting poetry, they don't need to determine how many syllables are in each sentence, but the pianist needs to know if they are in 3/4, 4/4, or even 7/8 time before they hit the first key. The musician is also required to do math on the fly, dividing beats by quarters, thirds, and sixteenth as they simultaneously map the notes onto the keys or strings of their instrument. Tempo markings and dynamics are a must for the reading musician, but the reader of words is free to deliver them however they choose (although for public speakers, maybe marking their speech "andante" or "mezzoforte" would be welcomed by their listeners).

How do music readers do it then? The most significant factor that makes one musician a better reader than another is the ability to spot patterns in the written score, using an understanding of the harmonic structure of the piece to prime their recognition of expected groups of notes. This ability to spot musical patterns is neurologically distinct from the ability to recognize human faces, letters, or letter strings.

Good readers also look ahead, reading notes .7 to 1.3 seconds ahead of what their fingers play, termed "eye-hand span". The musical pro sees four notes ahead of what they play, while the amateur looks only two notes into the future.

Seeing what is ahead is not as simple as gradually keeping the eyes ahead of the hands. When watching anything of interest the eyes normally perform saccades, quick back and forth (or up and down) eye movements, typically 20-30 times a second. This allows the part of the retina with the highest resolution, the fovea, to scan back and forth over the area of interest, leaving it up to the brain to merge all of these visual sweeps into a coherent picture. As one might expect, expert sight-readers have more efficient saccade movements, scanning farther forward into the score, and then returning to the position that is being played, dozens of times every second. The number of saccades is not related to the number of notes in the music, however.

Reading music is challenging, what with all that pattern recognition, looking ahead of what your hands are doing, and with those ocular saccades bouncing back and forth. Time to reconsider whether reading those black dots that musicians often term "fly poop" is worth it, when we just want to play "Orange Blossom Special".

Let's turn to wisdom from a board-certified neurologist: Josh Turkett, MD is a cognitive scientist, but he also founded the Brainjo Academy and wrote "The Laws Of Brainjo", a small but fascinating book on how to use the most current brain science to become proficient at the banjo. I know, it sounds like the beginning of a joke: "A neuroscientist and a banjo player walk into a bar…", but Dr Turkett offers a trove of ideas for learning banjo—or even a real instrument. Although he admits that reading tablature (the pictographic system of string music notation, see below) can be helpful to figure out just what Earl Scruggs was doing on that bluegrass tune, he is mostly against using written music. And he should know; after all, he is a brain scientist. All learning involves making neural connections, he emphasizes. When we learn to turn written symbols into musical motion, over time we create print-to-motor networks in the brain. But for most of us the goal is to create sound-to-motor networks, imagining what the music sounds like and having it emerge from our instruments. Turkett quotes Alfred Koryzbsky, guru of general semantics: "The map is not the territory". One of the "Laws of Brainjo" is that you can use notation occasionally, but you should get your eyes off the page as soon as you can to strengthen the sound-to-motor pathways. Even classical musicians could use this suggestion, memorizing complex pieces as soon as they are able, and maybe spending the some time at Fiddle Camp.

Brainy banjo players aside, composers continue to find new ways to put their ideas on paper (or now on iPads) for musicians to translate. For beginners, color-coded systems have been developed, using colors that match the colored keyboard notes to indicate pitch, lattices to indicate duration, and note sizes to show how loudly to play. This system won't get you to play "The Entertainer", but for the 5-year-old trying to match page to note, it is easy to understand.

There are any number of attempted improvements on the standard notation. Several, such as Dodici, ClaireNote, TwinNote, and Dodeka, use a larger number of staff lines and notes that straddle the line, sit on the line, sit between the lines, or sit under the next line, which expands the number of different notes that the staff can hold. The common factor is that these systems do not use sharps or flats; all 12 tones have their own position within the staff, eliminating the need for a key signature and accidentals. Since most Western music is tonally centered (the key of D major with 2 sharps, etc), these egalitarian notation systems might pull composers away from key-centered music. Whether that would be a good thing would be up to the audiences. There are differences between these four notation systems: Dodici uses little round o's for notes, ClairNote uses ovals whose length can be stretched to indicate how long the note is held, Dodeka uses rectangles that can also be stretched, and TwinNote mixes it up by using triangles, some pointing up and some down, some open and some filled in.

Musicologists have designed other notation systems to provide the information needed to turn heard music into played music. The long list includes Chro-Nat, Hass, Hellenic, I-accord, Klavar, Meloz, Symmetrisch Chromatische, and NUME (New Understanding of Musical Expression). These and many others are primarily intended for Western instruments and compositions; other musical cultures, such as Indian ragas, use their own systems, such as Svara.

What all of these systems have in common is an attempt to give the musician the information needed to reproduce the music, but choices have to be made as to what information is needed: Pitches? Duration? Articulation? Volume? Harmony? Feeling?

A Chinese woman used to hold jam sessions that I attended every Saturday at her house on Seattle's Capitol Hill, where she invited good amateurs to play jazz while she sang the songs of Gershwin and Porter. One weekend she had a woman friend visiting from mainland China, who politely listened to a few tunes. After Jack, a pianist with lots of technique and a sophistical harmonic vocabulary, had played a particularly complicated solo over "I Got Rhythm", the visitor approached the piano, pointed at his music, and asked in somewhat broken English, "Where are all those notes you played?" As is customary for jazz musicians, he was using a one-page "lead sheet" that provided only the basic melody and the standard chords. That was enough information for Jack to play the tune and two choruses of improvisation using the chord progression of the song, adding substitute harmonies and scales as he explored the form. He didn't need much information to make a lot of music.

Guitar players have their own thing; most of them can't read. Yet guitarists (and banjo players) have developed their own alternative notation, termed tablature. There are 6 lines, corresponding to the six strings of the guitar, and each note is represented by the number of the fret played on the appropriate string; an E note might be designated with a 0 on the top line which represents the first string, A as a 2 on the third line to show the second fret on the third string. But E might also be written with a 5 on the second line, or 9 on the 3rd line—all the same note, but on guitar the same pitch can be played on different strings. Literate musicians tend to look down their noses on guitarists who use tablature (referred to as Tab), but some information can't be indicated as easily in standard notation; in a complicated classical guitar piece, for instance, it can be very helpful to know upon which string each note in a 4-note voicing is played. Tab might be considered cheating by the modern Julliard graduate, but organists used a form of tablature in medieval times, and Renaissance lute players were using tab even as standard notation was still being developed.

Different strokes are needed by different folks to provide the information needed to produce music. In country music, the Nashville Number System (NNS) indicates the harmonies and rhythm, leaving the melody to the singers and the fiddlers (who were playing by ear anyway). Numbers are used for the scale step that a chord is based on, regardless of the actual key of the song. If the singer is in the key of C, a 4 indicates an F chord, and additions can be added to denote the flavor of the chord—D minor would be 2m, A minor 7 would be written 6m7.

Articulation might be indicated by a diamond shape around a chord that is to be allowed to ring, a dot represents that the note is played staccato, and a line under the chord tells the musician how long the chord should last. Enough information, the rest is up to the singers and fiddlers.

Finally, the tables may be turned. Instead of music being represented by ink on paper, the paper images might come first, and it is up to the musician to find their own musical expression of the composer's visual images. The score for Cilla McQueen's "Picnic" features four hand-drawn horizontal lines, one each for violin, oboe, bass, and guitar. Splashes of colored strokes are painted across each line to suggest sounds or feelings that the musicians should express, and it is up to the player to figure out how they can be turned into music. Less information, more inspiration.

Pianist Dave Brubeck grew up on a cattle ranch in Northern California and attended the College of the Pacific, intending to study veterinary science. His heart was in music, however, so he transferred to the music department. There was one problem: He couldn't read music at all. He was able to hide this deficiency by taking general music courses and playing cello and clarinet, learning enough scales and melodies by ear to get by. When he was a senior, the dean insisted that he had to take piano, since he was a pianist. In an interview with NPR radio, Brubeck told how a teacher discovered that he couldn't read and immediately reported it to the dean, who called him a disgrace to the conservatory and told him that he could not graduate. When the younger teachers heard this, they told the dean "You are making a big mistake because he writes the best counterpoint that (we've) ever had." The dean relented, as long as Brubeck promised never to teach and embarrass the conservatory.

Dave Brubeck's famous album of jazz in unusual time signatures, "Time Out", was inducted into the Grammy Hall of Fame in 2009 as the first jazz album to sell a million copies, and was certified as double platinum. He couldn't read, but it apparently didn't hurt his playing.

Touching Me, Touching You: haptics in music

From the first note, you know who is playing. The piano keys sounds like they are being stabbed by the musician's fingers, and it can only be Thelonious Monk. When Bill Evans plays "But Beautiful" on the same piano, each note sounds completely different, and not just because of the melody or harmony.

During the 20th century, classical piano was defined by two opposites: Glenn Gould, who hovered low over the keyboard to coax his delicate and exact sound, and Vladimir Horowitz, the Russian virtuoso who attacked the keys as if Napoleon was at the gates. Pianists' personalities and emotions can be heard in the way they press the piano keys; they call it "touch".

Piano teachers spend endless hours trying to instruct their students on how they should angle their fingers before touching the keys, how to feel the ivory and ebony (well, now that would be white plastic and some hardwood strips stained black) with their fingertips, and how to picture their fingers dropping on the keys by gravity or being pressed down with the flexor tendons. The full range of musical emotions cannot be expressed without the complete control of fingers on keys, and passionate listeners will rave about the subtleties of one pianist's touch over another.

And yet, physicists maintain that the only variable that the pianist can control is the velocity with which the felt-covered hammer activated by the key lever hits the piano string. Whether the key is slapped with a flattened digit or pressed by the tip of a gently curved finger, the only thing that matters is how fast the hammer hits the string. That's not what your piano teacher will tell you, however.

In order to test whether different ways of touching the keys actually produce different sounds, researchers recorded expert pianists playing the instrument in two different ways: Pressing the key (pushing down the key starting with the finger already in contact) or dropping the finger onto the key. The volume, that is the velocity of the hammer hitting the string, was measured and kept constant. When piano teachers listened to the two different touches on the piano, most could guess whether the keys were pressed or struck. However, when the sound of the finger hitting the key surface and the sound of the key reaching the bottom of the keyboard frame were digitally removed, the teachers guessed no better than chance. The conclusion was that the only difference between the two types of touch was the mechanical noise of the piano, not the musical tones produced by the strings.

And yet, the idea of touch persists. Perhaps the interaction between the musician and instrument travels both ways; the pianist plays, but also reacts to the feel and physical feedback of the keys while playing. A gentle caress of the keys in a Chopin nocturne causes the pianist to linger over the note a millisecond longer, while the forceful thundering of a Rachmaninoff concerto encourages the fingers to strike the keys again more aggressively.

The influence of touch extends into the listeners as well. An audience member watching a piano virtuoso playing Beethoven, raising his hands high above the keys before pounding a chord, perceives more animation, more urgency, and the music sounds different. Performer and audience interact with each other to combine the sound of the piano strings with visual perceptions. Pianist Glenn Gould is always mentioned when the subject of touch is raised; he would often play a sustained note, vibrating his hand back and forth above the key after it had been played, and his listeners insist that they could hear a vibrato that it created, even though there was no physical contact.

Touch is actually two different things, connected in the brain. The skin of the fingers have several types of mechanoreceptors, nerve endings which respond to forces in the environment. The sensors each have specific functions and charming names: Pacinian corpuscles for pressure, Meissner's corpuscles for touch sensation, Merkel's disks for vibration, and Ruffini's corpuscles for tension in the skin. In addition, nerve endings in the muscles and joints report the stretching and position of the hands, limbs, and torso. All of these receptors send their reports to the brain to provide up-to-the-millisecond information about the interaction of the body with the instrument.

Touch also refers to the outgoing signals from the motor control centers in the brain, instructing the body, limbs, and digits with high-definition instructions about where, how, and when to press the keys. Together, the touch-as-sensation and the touch-as-instructed-movement form a circuit within the central nervous system to give a performance of "Claire de Lune" its aura of wistful nostalgia.

Even the fingertips themselves have their own memory to add. Recent discoveries suggest an effect of "viscoelasticity"; after the fleshy tip of the finger makes contact, the spongy tissue returns to its original position slowly, and the memory of past touch is sent to the brain along with the report of the current forces.

In most areas of the body we become less sensitive to touch with repetition, termed habituation. But the fingertips of a pianist become more exquisitely sensitive with practice. Not only do musicians develop more fingertip sensitivity the longer they play their instrument, but their sense of touch is greater after they have practiced for a few hours than when they sat down to begin. Studies conclude that "Repeated execution of a tactile task enhances task performance."

If musical touch at the piano is much discussed, it is even more complex in instruments that can hold a constant note. The touch of a horn player's lips controls everything, and the feedback loop from brain to touch receptors in the lips and back is critical for intonation, tone, and attack.

No other instrument uses touch as much as violin. The left hand feels the balance of the instrument and the position of the hand along the smooth maple of the neck, while the fingers feel for the strings; each feels a little different, the low G string a thin rope under the fingers, the high E string a

thin live wire. But the pinnacle of musical touch-sense is the right hand, with its thumb and fingers wrapped around the base of a flexible two-foot-long stick of wood with horsehair stretched along its length.
The same touch receptors of a pianist's fingers have a dozen more functions on the violin bow. Teachers will tell their students that tone and volume are controlled by the speed with which the bow is drawn across the strings and the weight, or "bow pressure", that the arm exerts to push the bow down on the strings. But the sensation of starting a note, with the rosin-coated horsehair grabbing the string just before the bow begins to travel, takes years to master. The right arm pulls the bow to produce sound while feeling the vibration of the string thru the bow hair, thru the flexible wooden stick, thru the frog of the bow (inlaid with abalone, with its own smooth feel), and into the tips of the thumb and fingers wrapped perfectly—not too tight, not too loose—around the base of the bow. With so much information presented to the fingers thru the wood of the bow, it is not surprising that many bows cost as much as a late-model car. Carbon fiber and select other exotic hardwoods are used for bows, yet the gold standard is wood from the Brazilian Pernambuco tree, an endangered tree in an endangered coastal rainforest habitat. Touch is a precious commodity.

As music listeners, we are touched by the vibrations. We can feel the electric bass at a rock concert or the low brass in Wagner's "Ride of the Valkyries" with our whole body, like getting a shoulder bump to the chest with each note. We likely sense some vibrations with other parts of the body as well, but mostly with the tympanic membrane, the thin slice of tissue that we call the "ear drum". This works well enough when the array of nerve cells attached to the tympanum are healthy, but people with hearing loss are deprived of enjoying music. There are, however, different ways of hearing music.

Evelyn Gwennie is a celebrity for being deaf, but famous for being a musician. Her story has been told hundreds of times in print, TV, and film, but she is frustrated that the media are only interested in how a deaf person could possibly play music. And yet, Gwennie is one of the foremost percussionists in the world, playing with first-rate orchestras, composing, and collaborating with artists as diverse as Bjork, Bela Fleck, and Bobby McFerrin. She contributed to the spectacular opening music for the 2012 Olympics. Scottish-born Ms. Gwennie started losing her hearing at 8 years old, and by 12 she was profoundly deaf; that is, her ears could barely perceive some vibrations, but not enough to make out speech or distinguish the ringing of a telephone from background static. But, she insists, that doesn't mean she can't hear. She noticed that she could feel different vibrations on different parts of her body; very low notes were palpable in her legs and feet, while higher pitches could be felt on her upper body and face. Working with her teacher, she practiced distinguishing different sounds by touching surfaces around her. She notes that very low vibrations, such as a truck passing on the roadway, can be felt more acutely by the body than by the ears, and even though higher pitches are not as easily perceived by the body, she suggests that the feelings are there, simply washed out by the intensity of input from the ears. In addition, the sounds that we perceive combine sensations from the ears, the body, and the eyes; she watches other musicians (and can even visually discriminate different pitch vibrations of the typani), feels the floor (often playing in bare feet), and responds to the music all over her body. "Hearing is basically a specialized form of touch", she reminds us in her essay on Hearing found on her website, and the brain simply interprets all of the sensory input without having to think about it. "I have no more idea of how I hear than you do", says Evelyn Gwennie. All we need to do is listen.

Maybe haptics can help.

Cel phone users are already familiar with haptics; when your phone rings, it also vibrates, and even if the phone is silenced the vibration in your pocket tells you that someone is calling. The same basic concept is being developed to allow the deaf to "hear" music, but it is complicated. Our ear contains 10,000 hairlike sensors that respond to different pitches, and whose neural outputs are processed in the auditory cortex to produce the music that moves our emotions and our bodies. If only we could encode all of the vibrational information that we feel on our skin, we could approach something like hearing music; not with the sensitivity of Evenlyn Gwennie, but enough to pump our fists to Queen's "We Will Rock You", or feel the struggle between dark and light in Beethoven's "Ninth Symphony". With this in mind, the electrical wizards at MIT decided to try vibrating anything they could think of. The simplest haptics involve a tiny microphone and a vibrating patch applied to the skin of the forearm, programmed to buzz at the same frequency as the signal that the microphone senses. This conveys too little information, since only one frequency can be transmitted at a time; enough to suggest a pure tone but not enough to capture harmony or timbre, and limited to a practical frequency range of 65Hz to 1047Hz (four octaves in the lower and middle part of the piano keyboard. Further development added more vibrators spaced on skin around the body and programmed to respond to different aspects of heard music. Surprisingly the brain is able to receive all of the skin sensations and route them to the auditory cortex, where input from the ears is normally processed. One team designed a wearable vibrotactile vest programmed to translate sounds into a complex symphony of dermal stimulation.

Originally haptics were designed for the hearing impaired, but adventurous engineers started looking for ways to enhance the musical experience for people with normal hearing, a truly surround-sound experience.

Volunteers who have tried supplementary haptic devices report that there is a fine line between being able to appreciate subtle vibrations on the skin and being annoyed by the constant buzzing. Haptics may not be ready for prime time with concert-goers or at-home audiophiles, but stay tuned.

Musicians and teachers have a different idea for the use of musical haptics. Why not design instruments that guide the experience of the musicians themselves? Sensors could provide feedback to let the music student feel when they play the correct note or when just the right amount of pressure is applied to the digital piano key.

If haptics can allow the deaf to hear and enhance the perception of music for the hearing-able, the next logical step might be to eliminate the actual performance of the music altogether. If we could discover what elements of music move our emotions, what we find beautiful about vibrations in the air, we could program those aesthetics into haptics that could be enjoyed for their own sake. Musical principles such as repetition, rising and falling patterns of frequency and amplitude, or alternation of silence with stimulation might all be used to create their own form of art, going far beyond the limitations of 88 piano keys or 4 violin strings.

A team at MIT set as their goal the "composition and perception of intricate, musically structured spatio-temporal patterns of vibration on the surface of the body, a step toward an aesthetic of touch." The vibrations applied to the skin were not designed as information about an external event, but as interesting artifacts on their own.

As development proceeded on the project, a variety of different vibrational transducers were tried (this was MIT after all, so there was not shortage of cool electronic devices). The first prototype was the V1220, produced for use in tactile hearing aids. Once the engineers realized that low-frequency stimulation was essential to the vibrotactile experience, they added the Aura Interactor, a system designed for gamers that uses a wearable backpack to produce a sort of body-woofer effect. Three transducers were positioned equally along each limb, and the low-frequency vibrator was pressed against the lower back. Each of the transducers was controlled by separate outputs from the Protools Digi001 Interface.

The score used to control this symphony of touch was both less than musical, and more. The composition was done with Protools, a MIDI program used in recording studios, but the actual structuring of the result depended upon rethinking the parameters to be used. Normally the frequencies (pitches) of musical notes are of primary importance, but the skin doesn't perceive frequencies as well as the ear, so pitch became less important. Intensity, however, is easily programmed and perceived, so crescendo and diminuendo, sudden sharp attack, or gradual fading could be shaped to give contour to the vibrotactile experience. Texture could be added by using either rounded sine waves or rough square waves (or combinations), which the skin is able to distinguish easily.

With the hardware in place, the software programmed, and a structured sequence of vibrational events completed, it was time to performance-test the concept of music-without-music.

In September of 2001 a series of experimental events, dubbed Cutaneous Grooves, were held at the MIT Media Lab. Each audience member donned a full-body vibrotactile stimulator suit and sat for a one hour event. The "warmup" set was composed of audio sounds of heartbeats and blood flowing thru the arteries, followed by a segment of tactile stimulation, morphing into an audio and tactile thunderstorm.

The main event started with layered percussion and a haptic interpretation of the audio, so that the audience could try to attend to both streams separately as the audio and tactile compositions alternately came to the forefront. At first the audience members weren't sure what to think or feel, but it didn't take long to notice repeating patterns in the vibrations and show surprise when the patterns were broken. When the accompanying music was calm, the tactile stimulation was also felt to be soothing, and when the music turned strident, the vibrations seemed harsh, almost painful. The experiment confirmed that sensations on the skin could enhance, and perhaps take the place of, musical sounds. After all, "hearing is basically a specialized version touch".

Ludwig von Beethoven started losing his hearing in his twenties and could no longer hear by the time he was 44.

Numerous causes have been proposed, but the tragedy of a musical genius who lost the ability to hear music is one of the great human stories. His later symphonies, composed in total deafness, stand as a testament to the resilience of the human spirit. The household servants that cared for Beethoven late in his life reported that they often saw him composing at the piano with one end of a long pencil held between his lips, the other end resting on the piano soundboard to feel the vibrations of the notes he played. When hearing fails, touch may provide its own comfort.

The Long and Winding Road: the voice

I can't sing. Well, technically that isn't true; I can produce vocal sounds that more or less match a pitch that I hear or imagine, but nobody would want to hear the sounds produced. It isn't just me, either; for years I played guitar and led a contemporary choir at our church and sang along with the other members of the group that led songs at mass. Until our priest called me in to make a request: "Please don't sing along with the choir."

In my essays about music and the brain, the art of singing is noticeably absent. I prefer to listen to music without lyrics. And yet, the voice is the primeval musical instrument, and all the constructions of wood and wire, metal and wind, are poor imitators of the vocalizations that seem in many ways instinctive, effortless. Naturally, I am curious about how the voice can create music that either sounds sublime, or simply hurts the listener's ears.

When I was born, the pediatrician who examined my mouth declared to my parents that I was destined to be a good singer; something about the arch of my mouth or the shape of my uvula or something. My parents believed him, and I grew up expecting that I would develop a wonderful voice. I didn't actually sing to test this theory, but when I was 14 I took up the guitar, thinking that it would enhance my eventual singing ability. I quickly became competent on guitar, but then discovered that my singing voice would only be useful as a wildlife repellant. From then on I focused on music without words.

Since this book is about how music is created by the central nervous system, I feel compelled to dissect the brain to map the voice, but perhaps this is too challenging a task.

With the advent of advanced brain imaging like fMRI and PET scans, scientists can identify the spots that are active during singing, but enumerating the locations involved is like describing a road trip from Chicago to Boston by simply listing the towns along the way; you know where you have been on the map, but miss all of the scenery along the way. Singing touches on more parts of the brain than any other human activity.

Almost all of the major parts of the nervous system get involved, whether an opera singer is performing the "Habanera" aria from Bizet's Carmen or Mick Jagger isn't "Getting No Satisfaction": The frontal, temporal, and parietal lobes of the brain are involved, along with the cerebellum (usually thought of as the center for muscular coordination) and the brainstem (where many of the automatic functions that keep the body running are located). The insula and cingulate cortex are critical, especially in passing the melody, harmony, and rhythm back and forth between nerve centers in opposite cerebral hemispheres. Broca's area, famous for its role in language, is involved as well.

The small towns along this road trip include the precentral operculum and gyri, the anterior middle temporal gyrus, the superior temporal sulcus, the nucleus acumbens, the gyri of Herschl, and, of course, the Rolandic cortices. Numerous neighborhoods and crossroads are met on the journey, as the emotions travel thru the amygdala,and hippocampus, while rhythm crosses the pars orbitalis, insula, and left cingulate. Melodic repeats and harmony contribute their elaborations via the planum polare. The fact that most human brains can manage to carry a tune at all is astounding.

The right and left sides of the brain cooperate to merge their musical contributions, the wholistic and artistic right hemisphere, advised and analyzed by the verbal left cerebrum.

The asymmetry of musical loci is central to one well-studied phenomena: stroke patients who have extensive damage in Broca's language region in the left cerebral hemisphere lose the ability to talk (termed aphasia), but not to sing. Over a century ago, doctors noticed that some stroke patients who could not speak were able to sing the lyrics to familiar songs. This influenced neurologists in the late 19th century to ponder how the brain produces music, and during the current century it has spawned the development of singing-based therapies for aphasia, including Melodic Intonation Therapy (MIT). The theory behind MIT is that singing encourages undamaged areas in the right side of the brain, native territory to music, to take over the functions of the left side language centers.

In one typical case a woman suffered a severe left hemisphere stroke at 48 years of age, losing both speech and use of much of the right side of her body. After rehabilitation and intensive speech therapy she failed to show improvement, but after singing therapy was started she found that she could communicate with words and phrases sung as music, although her voice sounded much different. Similar cases of brain damage provide an opportunity to understand what happens when music unlocks some language ability. Imaging studies have given varied results; in some cases the musical right brain seems to gain some language ability (particularly in trained musicians, who spend more time in their right mind), but more recent tests show that much of the improvement is actually credited to development in the remaining areas adjacent to the left-brain damage. Many studies asked patients to sing lyrics from familiar songs or to speak them; singing the words to well-known songs was easier, perhaps because the lyrics were from songs that the person had known their whole life, and may have been stored in a deeper and more distributed manner than everyday usable language. In some trials rhythm was used, rather than melody; instead of singing a phrase the patients were asked to tap their finger in rhythm while speaking the words. Adding the rhythm alone produced as much speech improvement as did singing.

How is it that singing and speaking are so similar, and yet so different? While there is some degree of laterality (singing, right hemisphere, speaking, left hemisphere), it fails to explain the differences between speech and song. Enter the "dual stream" concept: While music and speech are both are supported by left-sided language networks, the output divides into dorsal and ventral streams in the brain. The dorsal stream is more general-purpose, carrying speech and music and providing coarse-grained pitch processing, while the ventral stream is unique to singing and includes more fine-grained control of pitch. Some researchers suggest that even though both pathways run thru the left brain, the ventral stream is in a less vulnerable location and accounts for the ability of some aphasics to sing lyrics. Other neural centers add their own flavors to the song; some areas keep an eye out for violations of the rules of speech, while other areas listen for harmonic mistakes, and both are added like tributaries as the dual streams progress.

Studies of the anatomy of music in the brain have been influential in our concept of how the brain works: The central nervous system can be understood as a complex modular system, each location adding, subtracting, or connecting in order to produce the pinnacle of human art.

Admittedly, as a non-singer who plays a number of stringed instruments, I would like to know if these many parts of the singing brain have anything to do with instrumental music. Fortunately, curious neuroscientists decided to test how singing and cello playing look in the MRI machine.

Cello was chosen because it demands the same auditory networks as singing; each requires the musician to make minute adjustments to keep their note in tune with a reference note. Cellists were asked to play a note in tune with a note played for them, and then the same subjects were asked to sing the reference note. Findings showed that the same brain pathways were activated, whether the voice or the cello produced the tone. The neural pathways diverged only where the motor output went to the hands or to the larynx. The most surprising discovery was that the output channels went to the larynx, even if the cello was being played and the larynx was voiceless. This effect was most noticeable when the cellist had started playing the instrument by seven years of age. For those early starters, the cello had truly become their voice.

The breadth of brain stimulation involved with singing has made it a natural choice for music therapy, particularly in developmental and degenerative brain conditions. Aphasia was discussed above, but stimulating a variety of modules within the central nervous system has shown promise in other areas. Parkinson's disease affects almost a million Americans, mostly over 60 years old, and causes a devastating lack of control over movement, breathing, swallowing, and control of the voice. Singing, particularly in a group setting, can improve the intelligibility of the Parkinsonian's speaking voice. Walking gait can be improved with rhythmic aids, like a metronome, but singing while walking shows more improvement. Perhaps most importantly, Parkinsonians suffer from frustration at not being able to control their movements, and group singing with others who are similarly afflicted lifts the cloud of depression that comes with the disease. In every study, no matter what the measurable neurologic outcomes were, the participants said that it was enjoyable, and more justification is hardly needed.

The benefits of singing have been shown with other neurologic conditions. Dementia patients connect with familiar songs of their youth, and the stimulation of so many parts of the brain slows the forgetting and confusion. Singing, whether alone or with a group, increases connections in the brain and lifts the mood, even when there is much to be melancholy about.

If singing benefits the person who can't talk, move, or remember, it is no surprise that there are benefits for the healthy, normal individual (if there are truly any "healthy, normal individuals" among us). One extensive review confirmed that vocal music makes us breathe better and offers cardiovascular benefits. One typical study of 210 adults concluded that "singing has been shown to help people ease anxiety, brings euphoria, and creates a sense of belonging." Many of these studies also pointed to increased neuroplasticity in the many parts of the brain along the singing pathways. With all these benefits, in the words of the 19th century American hymn, "How can I keep from singing?" For some of us, it is by popular request that we refrain from singing the refrain. When our priest asked me not to sing with the choir that I led, I was determined to learn the secret of creating listenable sounds with my voice, and I studied with a singing teacher for a year and a half. For the first month, Ms. Stanley forbid me from attempting to make any sound at all; first we had to train the muscles of the throat to make room for the music to come out, she told me. For an hour each morning I would do exercises that involved putting my index finger down my throat as far as I could without gagging, to gradually relax the muscles in my throat so that they didn't constrict the flow of sound. It took a while, but eventually I could reach down my throat as far as my finger would go without causing myself to vomit (while an accomplishment, I have yet to find any other use for the skill). Then I used a flat board with a concave arch on one end, which I pushed against my throat from the outside while pushing forward and down with the muscles below my chin, sort like of neck pushups. I gained strength and the ability to expand my throat voluntarily, and with my newly developed muscles I was finally ready to make sounds. The results were disappointing. I learned a lot about how the singing voice works, but still had to confine my singing to lullabies for our children (when they were too young to complain).

All of this brain science and practice should have allowed me to find my own singing voice, but the truth is that I just don't sound very good. I can hear the notes accurately, I can match the pitches of the music, but that isn't enough to create a pleasant sound. As far as I know, my brain's temporal lobes, insula, and cingulate cortex work just fine. Even my cello playing is reasonably in tune, so my problem seems to be more in the motor output side of vocal music production.

So far I have neglected the output part of this story. Once the brain has done all of its magic of hearing, analyzing, and planning, the resulting product needs to be sent out to actually produce the sound. Of course! The beautiful sounds I hear in my head are just not being delivered in the way that I imagine them! In what seems like an improbable and inefficient way to send a song to the body's "speaker" in the larynx, the vagus nerve is called upon. The vagus nerve (named for its wandering course to various organs around the body), leaves the head thru an opening at each side of the base of the skull, with branches that carry messages to and from the intestine, liver, heart, and lungs. Several branches of the nerve control the vocal cords that produce the voice, but strangely (except for one small branch) the nerves don't go directly from the base of the skull to the nearby larynx. Instead, the fragile branches of the vagus that give us our voice travel down the neck alongside the carotid arteries, and then loop around the blood vessels at the base of the heart before turning around to travel all the way back up the neck to the voice box. This down-and-back route earns this branch its name, the Recurrent Laryngeal Nerve. Because of its circuitous route, the Recurrent Laryngeal Nerve is the longest single nerve in the body, which makes it vulnerable to all sorts of injuries, malfunctions, and deteriorations from age or neurologic disease. When we listen to the voice of an aging Sinatra or Caruso, we can easily imagine that the journey down the neck and back is starting to become rutted and worn.

We all welcome a handy excuse for our shortcomings, so with no medical confirmation whatsoever I will blame the recurrent laryngeal branch of my vagus nerve for failing to deliver my imagined song. "The Long and Winding Road", indeed.

The Beat Goes On: binaural and isochronic beats

When I was 15 years old and trying to learn how to tune a guitar, my father showed me a neat trick. I don't remember a lot of his valuable fatherly advice, but this one stayed with me.

Dad had been a musician in his younger days, playing trumpet, sax, and some boogie-woogie piano. He even had his own big band in Milwaukee when he was 18, but at 19 he was in the infantry, shipped out to France to fight in WWII. After he suffered a head injury, returned to the States, and married my mother, he changed course and became a physicist. As a scientist that specialized in electronically measuring strain, he worked on everything from the Titan missile to the massive antennas that monitored Russia during the Cold War. Dad never missed an opportunity to explain any natural phenomenon, at length—great length!

"Listen to this.." Dad said as he played a low E on the piano. "Now play the low E string on your guitar. Can you hear those beats?" I listened carefully, but all I could hear was that my guitar string was not exactly in tune with piano; I couldn't tell if it was flat or sharp. "Listen again; if the two notes are close but not perfectly in tune, you can hear a beating sound, wa wa wa wa." I listened again and there were the pulsations. "Now tune your guitar string slowly and listen." I did as he said, and I could gradually hear the pulsations becoming less frequent, until they disappeared altogether. "There, now that string is in tune with the piano". And of course, an explanation followed.

"Sound waves are just pressure waves in the air, and like any wave they have a high peak and a low trough. In the case of the low E note on piano, there are 82.41 waves per second, or 82.41 Hertz (Hz). If your guitar string is a little flat, say 81Hz, then the waves won't line up exactly. When the wave of your guitar string is going down and the wave from the piano note is going up, their volumes will cancel each other. Once they catch up and both waves hit at their peak together, the volume will be louder. These are the beats, or periodic volume modulations, that you hear. The closer the two tones are, the slower the beats, and once the beats disappear then you are in tune." Simple physics. Thomas Young recognized the phenomena in 1801, and it is still true.

Sometimes the interference of nearly identical sound waves is pleasant: It is the sound of an old piano that has been sitting in the basement and hasn't been tuned in several decades. Because each note on the piano involves a hammer hitting two or three strings of the same pitch, being slightly out of tune creates a rippling sound like flowing water. While this might drive the classical pianist crazy, it has a certain honky-tonk charm. Twelve-string guitars naturally have a little of this in-and-out-of-phase sound, since you can never keep the pairs of strings exactly in tune with each other. Electric guitarists often run their instrument thru a phase shifter pedal, which divides the guitar signal into two waves and delays one of them so that there is a rhythmic interference to create a similar sound.

Waves of sound give us music, but there are other waves with interesting effects. With the development of the electro-encephalograph, neuroscientists are able to measure rhythmic oscillations in the brain, millions of neurons firing in the same rhythms. Not only does this open new areas of research, but popular science, pseudo-science, and New Age fantasy have embraced the idea of listening in on brain waves and controlling them for health and wellness. By matching the frequency of the oscillations with how people reported that they felt and acted, a menu of desirable brainwave frequencies has been compiled, the same correlations mentioned in the previous chapter on meditation:

Delta, .5-4Hz associated with sleep
Theta 4-8Hz very relaxed and internally focused
Alpha 8-12Hz relaxed with passive attention
Beta 12-35Hz active, alert, and focused
Gamma 30-80Hz high levels of thought and concentration

The implications are intriguing to both scientists and spiritual seekers: If we could change a person's brainwave frequencies, we could finally control the brain (and who doesn't want to tell their brain what to do, instead of the other way around?). Enter the concept of entrainment. If a group of people are walking together, their strides will unconsciously sync up (or consciously lock steps, if you are a platoon of soldiers or a marching band). Entrainment is a common phenomenon in physical and biological systems, so it seems possible that the rhythms of the brain could fall into step with external stimuli. Scientists have used flashing lights at desired frequencies and demonstrated that the neurons can be coaxed into oscillating in sync with lights. Why not sound?

The auditory beats created by two nearly-in-tune musical tones could provide such a rhythmic stimulus. Theory says that if we want to attain the totally chilled-out state of theta waves (found in meditating Tibetan monks), we can can listen to auditory beats at 6Hz until we are suitably mellowed out, without all that bothersome sitting cross-legged on an uncomfortable floor. Fortunately, this effect can be objectively measured by recording with an EEG while the person listens to the beats produced by two nearly identical musical tones. An attractive theory, but studies have been ambivalent and contradictory in their results. In any case, listening to two slightly out-of-tune tones for 20 minutes may not be preferable to sitting crosslegged on an uncomfortable floor.

When the auditory beats are heard with both ears, they are termed "monaural beats", which is a little confusing. But at least the ears agree on what they are hearing. A more complicated version of beat entrainment is termed "binaural beats". When two pure-but-slightly-out-of-tune notes are heard by different ears, the auditory centers of the brain produce a third tone midway between the two frequencies, and this intermediate frequency shows amplitude modulation, or beats. These binaural beats are similar to monaural beats, but they come with a ghostly extra tone that exists only the in the brain. Binaural beats are produced at a frequency equal to the difference between the two original frequencies. Chilled-out theta beats can be produced by entrainment at 6Hz by exposing the right ear to a musical tone at 440Hz (A below middle C) and the left ear to a tone of 446Hz (an eighth of a musical step higher), creating the illusion of a tone at 443Hz that modulates in volume at 6Hz (as long as you listen thru earphones to get the audio separation). This unusual auditory illusion has spawned an off-beat recording industry of its own.

We will return to the question of whether all these beats actually cause entrainment of brain waves, and whether achieving an alpha state to relax, or a gamma vibe when you are taking a test is really useful. But first, let's go shopping.

Playing the part of curious observer, I Googled binaural beat recordings and found hundreds of CD's and what seemed like a never-ending variety of YouTube videos. With earphones in place I checked out as many YouTube clips as I could, but stopped listening after the first dozen, a little confused.

1. The simplest track provided a pair of solid tones with beats at 40Hz. I think I could hear the ghost tone, but if there was volume modulation at 40Hz it was too fast for me to make out.

2. The next sample I checked out sounded like a slow progression of notes (not single tones) with volume that swelled and faded twice a second—2Hz. I moved on before slow delta brain oscillations took over and I fell asleep.

3. Another recording had no audible beats, but the overall volume went down every few seconds and then came right back up. If this was supposed to be binaural beats, the frequency would have been .5Hz, low enough to suggest a coma. Since the volume

modulations were abrupt, I assume that it was just regular music with the volume turned down every few seconds.
4. One of videos showed a slight warbling audible at 6Hz, along with calming synthesizer music.
5. I listened carefully to a sample with a very deep bass tone accentuated by a periodic gong, accompanied by a heartbeat pulse—lub dup—every few seconds. Interesting, maybe even relaxing, but I couldn't pick out a binaural beat, just a heartbeat.
6. A sample of peaceful music used a drone, hard to tell if there was a binaural beat behind the keyboard arpeggios and repeated melodic figures that sounded like they were played on a Telecaster with a reverb pedal.
7. I made it as far as a track titled "528Hz Healing Frequency of Love", didn't feel any different, and lost interest. There was another recording intended for dogs, but I didn't check it out.

From what I can tell from the website blurbs and product claims, I suspect that most of the BB products generate binaural beats, but they are buried in a mix of other music, sort of like the "enhanced water" that claims added vitamins; you can't tell that they are there, but they might be good for you. In the theta and alpha frequency pure-tone tracks, the ghost tone with pulsing volume was audible, but it wasn't pleasant to listen to. In most of the "meditation music" tracks (including nature sounds like rain, as well as slow synthesized harmonies) the "carrier tones" that would produce the binaural beats were not obvious. I may be missing something, but if each note of the melody was combined with a corresponding tone in the other ear that was displaced by 6Hz to produce a theta binaural beat, the brain would not have a chance to lock into the beat as the carrier tone changed. If the binaural beats were simply a constant pulsing buried in the music, it wasn't apparent. But maybe that is the point: Our auditory processing ability might be able to pick out the pulsations hidden in the music so that our brain waves can match the oscillations.

Of course, most of these recordings are intended for purchase by people who want to hack their brainwaves, so we only have to trust that there really are hidden binaural beats. I did run across one scientific paper in which researchers designed an algorithm to determine whether a recording actually contained binaural beats or not (which makes one wonder about the binaural beat claims for most of the tracks). The study claimed that their technique to extract the binaural beat information was 95% accurate in identifying the phenomenon, but it didn't report how many of their test recordings actually confirmed binaural beats. Even if they could prove that a recording actually contained beats, it seems unlikely that anyone will ever offer a certification of testing for authentic beats. It all seems reminiscent of nutritional and herbal supplements, in which half of the products tested do not have the ingredients claimed on the label.

Sometimes we can think too hard about why something works, whether it is a nutritional supplement or a recording of trance-y wooden flute music over a low drone tone. I found listening to many of these tracks calming, and 20 minutes of sitting still and letting the sounds wash over me was a pleasant break from the world around me; whether my theta and alpha brainwaves were becoming entrained, I can't report.

Whether we can hear them or not, there is no shortage of beneficial effects that have been claimed for binaural beats.

According to one source, binaural beats can entrain the brainwaves and:

Reduce Anxiety

Increase focus

Lower stress

Increase sleep quality

Improve creativity

Enhance meditation

Create mindfulness

Bring positivity

Increase concentration

Bring body healing, and

Increase spiritual consciousness by "opening your third eye".

Another source tells us that BB will decrease fear and overthinking (who doesn't need that?)

Wonderful as all of these benefits sound, the question needs to be asked: Do binaural beats actually change the frequency of our neural oscillations?

As with all scientific studies, the results are variable, and can depend on the carrier frequency (binaural beats are perceptible when the tones that produce the illusion fall between 300 and 1500Hz), the frequency of the binaural beats themselves, the methods for measuring the oscillations in the EEG, and the self-reporting of the listener.

Positive results have been shown. When pure-tone binaural beats are used, there is evidence of brainwave entrainment, strongest at the alpha frequencies from 8-12 Hz with a 400Hz carrier tone (or a tone of 400Hz combined with tone of 410Hz in the other ear to give 10Hz binaural beats).

Browsing the scientific literature on PubMed, there are studies that seemed to show entrainment of brainwaves (also called Frequency Following Response, FFR); delta waves were used in one study, theta waves in several others. A number of tests couldn't identify brainwaves that resonated with the binaural beats, or confirm that alpha and beta beats caused a difference in the brain. It might also depend on the subjects: when experienced meditators were compared to those without meditation experience the meditators responded, while the novices did not; perhaps the experienced subjects were more used to controlling their brain waves and may have been more sensitive to the beats. Another study alternated matching pure tones with unmatched pairs, and showed that the brain response was greater during the matched tones, when no beats were present. It is notable that both white noise and music (with or without binaural tones imbedded) often made subjects calmer and more focused, suggesting that listening to the right sounds can have positive effects on the brain, whether the sounds are are exactly in tune or not.

If the rhythmic stimulus of binaural beats hidden in relaxing music can capture our brain waves and put us into a meditative (or alert and creative) frame of mind, perhaps other sounds can do it as well. This led to the development of isochronic tones. In this technique, bursts of a pure tone (not mismatched like binaural beats) are repeated at the frequencies chosen to produce entrainment with brain waves. If the calming effect of brain waves at 10Hz is desired, simply produce 10 on-and-off bursts of a tone every second and repeat for 10-20 minutes. The advantage over binaural beats is that earphones are not required, as the sound can be heard by both ears, similar to the monaural beats heard while tuning a guitar. It appears that the effect of isochronic beats is stronger than that of binaural beats, since the power and volume of the periodic pulses can be cranked up much higher and are not dependent on an auditory illusion. Studies seem to show the effects of entrainment more definitively with isochronic beats, but if listened to by themselves they quickly become boring or annoying. To avoid this effect, many of the isochronic recordings hide the tone pulses in other music, nature sounds, or white noise. That I could live with.

If both binaural beats and isochronic beats can be embedded in other music without being easily identified, one could wonder (if one were a conspiracy-theory kind of person) if we are having our brain waves manipulated without knowing it. The elevator music in a stock trading building on Wall Street could secretly pump out gamma beats to get their employees into a hyper-alert and aggressive mental state, while middle school classrooms could be filled with Taylor Swift tunes laced with isochronic beats at 10Hz to suppress the restless students. I will be listening for the beats whenever I hear music in public.

In the meantime, I think I will just go in and play the low E string on my guitar, slightly out of tune for 20 minutes. Seems like easier way to get my theta waves on than learning to meditate.

Hertz So Good: music at 432Hz

As the last stragglers find their seats, the orchestra starts to tune up. If it is a full orchestra, the oboe plays the first note, an A, and then the concertmaster (boss of the violins) picks it up and plays the note so that the rest of the orchestra can tune to it; conveniently, all of the stringed instruments have an A string, the first string on cello and viola, the second string on violin, the third string on double bass. I have always been mystified by how a violinist who is surrounded by twelve other musicians playing the same instrument can hear if their own strings are exactly in tune. Practice, I suppose, but I wonder if the tune-up is partly a ritual for show. It seems likely that most of the musicians tuned up with a cel phone app just before coming onstage anyway. However they do it, when musicians make music together, they need to be in tune.

To be "in tune" means that each note creates vibrations in the air at an exact frequency, but the frequency of that standard pitch is a subject of acrimonious debate. The current agreed-upon benchmark is that the note A vibrates at 440 Hertz (vibrations per second). But the choice of 440Hz is somewhat arbitrary, and it hasn't always been that way. If some people had their way, the modern standard would be repealed. The complicated story of standard pitch centers around the alternate frequency of 432Hz, and the plot of this story has everything: History, science, numerology, musical celebrities, conspiracies (either Nazis or a greedy musical instrument manufacturer, take your choice), New Age healing philosophy, and the vibrations of the Earth itself.

What's the matter with music tuned A=440Hz, anyway? With a little research on the internet I learned that music at 440 is less enjoyable than at 432, and that it has a negative effect on mood, consciousness, our natural energies, and our spirituality. It can cause a disconnect between body and consciousness and increase social friction. In fact, it seems to be responsible for most of the external and internal conflicts with which the modern world is rife.

Music based on A=432Hz tuning, on the other hand, relieves anxiety by synchronizing the entire brain and making it resonate with the natural elements. It increases our concentration, creativity, and spirituality. And as a tangible benefit, when 432Hz music was played for patients before dental surgery it decreased their heart rate, blood pressure, respiratory rate, and cortisol level. While I take break to re-tune my guitars, let's look back at the history of tuning frequencies, since many of the claims around 432Hz are based on the backstory of standardized tuning.

Let's go all the way back. Dr. Jelle Atema, a flutist and professor of Marine Biology at Boston University, gained access to three very old bone flutes and played them (or exact reconstructions) to investigate the musical frequencies of our distant ancestors. The oldest currently known flute was found in a Neanderthal cave dated 50,000 years ago, carved from the femur of a young cave bear. Playing a reconstructed version of this ancient flute, the frequencies corresponded to 466Hz, 493Hz 523Hz, and higher notes, none of which related to a multiple of 432. But these early musicians were not modern humans, and didn't have our advanced musical culture.

A more recent relic, the Flute of Abri Blanchard, is believed to be 25,000-35,000 years old. When played it produced frequencies of 369, 414, 466, and 493Hz. So far, no "Frequency of Nature."

One of the best candidates for hearing what very old music sounded like is the Flute of Veyreau, a beautifully carved instrument made from the wing bone of a Griffon Vulture, a mere 4000 years old. Dr. Atema analyzed the flute and built an exact replica, complete with the ornate carvings, from the wing bone of a modern Griffon Vulture (fortunately the species had only recently been rescued from the brink of extinction by a French conservation organization, and they managed to provide Atema with a suitable specimen of bone). The scale produced was not the familiar major scale that we all know, but the frequencies were 369Hz, 425Hz, 493Hz, plus some other notes, none in the vicinity of 432. But Dr. Atema reported that the flute played beautifully, with a clear and very controllable tone.

We now move up to the 17th century CE, when stringed instruments were being produced from trees that grew in the Alps during the previous ice age. Thank you, Mr. Stradivari and Mr. Guarneri. Unfortunately, we have little to learn about pitch from these old violins and cellos, since the strings can be tuned to whatever pitch one chooses, and pitch, like time, was all local. A string quartet in Paris might have tuned to A=412Hz, while a nobleman's court orchestra in Munich may have chosen A=490Hz.

Even with historic uncertainty, ancient music ensembles often choose to tune lower than modern orchestras, with the assumption that this was common in the old days when there was no standard pitch to which to adhere. In addition, older instruments may not have withstood the physical stress of strings cranked up to higher frequencies. Even these history-conscious ensembles don't agree on the ideal frequency for the A reference tone.

As it happens, we do have a few tuning forks used by the classic composers. Mozart's possessions include a tuning fork that produces A at 421.6Hz. Beethoven's personal tuning fork is preserved in the British Library, and a recording of its tone can be heard online, at 455.4Hz. So far, 432 doesn't seem to be rooted in history.

Galileo also dabbled with investigating musical pitch, but the first scientist to really dig into the virtues of music at different frequencies was Joseph Sauveur, a French acoustic physicist. In the late 1600s, Sauveur consulted musicians of his day and found frequencies for A that varied from 405Hz to 421Hz. In 1701 he proposed that a standard pitch be established that set C at 256Hz (which would correspond to 432Hz for A) because it made the math easier when studying vibrations in his physics laboratory. The musicians didn't like it and it never caught on.

During the next 200 years the pitches used by orchestras gradually crept higher. The construction of stringed instruments improved, so that they would tolerate greater tension on the strings, and the higher pitch gave a brighter and more brilliant sound, important as concert halls became larger and audiences wanted to hear more energy in the music.

Those who didn't like the higher pitch were the opera singers; as the notes were pushed higher, it put more strain on their voices. The Italian opera composer Giuseppe Verdi was a strong component of a lower pitch, and in the late 1800s he suggested 435Hz, and then 432Hz. His campaign never caught on either, but the proponents of the 432 standard still refer to it as Verdi tuning.

Despite the objections of opera singers and nostalgic string players, the International Organization for Standardization officially set the musical standard at 440Hz in 1955.

Enter Lyndon Larouche, an activist on both the far left and the far right at different times. When Larouche wasn't running for US president (which he did for every election from 1976 to 2004) or serving jail time for fraud, he founded the Schiller Institute, which lobbied in 1988 to change the standard tuning for classical music to 432Hz. He even used his influence in Italy to push for legislation to punish musicians for tuning at 440, and to confiscate their tuning forks. Larouche loved a good conspiracy, and claimed that the 440Hz standard was part of a Nazi plot to weaken people's minds and destroy music in the Western World. A competing conspiracy theory claims that in the late 1800's an instrument manufacturer named J C Deagan pushed to standardize 440Hz, and by 1917 had convinced the American Federation of Musicians to make the change. His motivation for the crusade supposedly was to sell new instruments to all the musicians who still had horns and woodwinds that were designed for the historically lower pitches.

Now the controversy lives on, and the cause is not without celebrity advocates. Opera stars like Placido Domingo, Luciana Pavarotti, and soprano Renata Tebaldi support the lower pitch (although one could wonder if their younger selves could have reached 440 more comfortably than their mature voices could). The idea has caught the attention of some pop and rock stars: Pink Floyd was said to have recorded in 432, and Prince mentioned the trend in a favorable light.

For the rest of modern recorded music, mostly performed at the standard frequency, there is an antidote. Digitally retuning recordings, from Handel to the Red Hot Chili Peppers, can be accomplished without slowing the tempo of the music. Listeners can download versions (when they aren't in copyright violation) of Mozart, Duke Ellington, Rush, Bruno Mars, and Ariana Grande re-tuned to 432Hz; some listeners report that they can immediately feel the difference and that they hear the music more clearly. They are calmer and more in tune with their true nature when they hear the 432Hz version. Music lovers can even download the "432Player", which can be used to play any recording at the lower frequency.

And of course, the 432 idea is perfect for New Age spirituality and meditation music, where the frequencies appear prominently on the music websites.

After all of this wandering around in the history and physics of the 432Hz music movement, where does that leave us?

First, few people would be able to tell the difference between the frequencies; If standard A is 440Hz, 432Hz would sit about halfway between A and G#, a quarter tone in music terms. Musicians often transpose music to different keys, and only a listener with absolute pitch could tell whether a song was played in A major or G# major, much less something in between. What about the concept of 432Hz being in tune with the resonance of the earth? The vibrations of the universe are only distantly related to any specific frequency, and efforts to make the two coincide seem like modern-day numerology. The claim is made that at 432Hz tuning all of the octaves of C are whole number frequencies, but we need to consider what Hz means: vibrations per second. Seconds have not always been the same, however, and A=432Hz would be A=440 if our modern standard second was 18 miliseconds longer. The duration of a second, while not arbitrary, is almost as convoluted as the musical frequency standard. The pre-scientific meaning of a second was "just a tiny sliver of time—I'll be back in a second." In the 1940's it was official that a second was 1/86400 of a mean solar day, but in the 1950s they realized that the earth's rotation wasn't consistent enough (are you listening, 432?) to use for really important measurements, such as high-speed photography or measuring Usain Bolt's time for the 100 yard dash. Now a second is scientifically defined as 1/31556925.9747 of the year 1900. The idea of the divine origin of mathematics and music is as old as Pythagorus (and 432Hz is sometimes referred to as Pythagorean tuning), but modern physicists seem unimpressed by the ancient Greek's "music of the spheres" vision of a perfect mathematical universe.

Theory is one thing, but the actual experience of music is another. Many enthusiasts of lower frequency music tell of life-changing emotional responses to 432Hz, so who can question their subjective reactions? Maybe the difference is all in the mind, but with the 432Player you can enjoy it however you want. Nobody is being harmed, so why not?

Curious, I listened to track of Jason Mraz singing "I'm Yours" at 432Hz, then the original recorded at A=440, then back to the 432Hz version; I think that I could tell there was a difference, but both had a chill vibe that made you just want to sit on a beach somewhere.

I also listened to several tracks of New Age and meditation music at 432Hz and they were—-relaxing, for what that is worth.

Scientific wisdom states that if you can't measure it, then you can't understand it, so I felt obligated to do a poorly designed study of my own. Since heart rate and blood pressure are two of the simplest measurements of how tense and stressed we are, I strapped on a blood pressure cuff, listened to Journey's "Don't Stop Believin'" at the original A=440Hz tuning, and noted my heart rate and blood pressure. Then I repeated the measurements while listening to the same song retuned to A=432Hz. There are obviously a lot of confounding factors in this quick-and-dirty experiment, which defies most of the rules of experimental science: The subject was not blinded as to which version he was listening to, time of day was varied (the experiment was done once in the morning while waiting for a doctor appointment, then repeated in the afternoon while watching a granddaughter jump off tall objects at her parkour gym), and the attention being paid to the music varied. At least the versions were alternated to eliminate habituation to the music. In this highly unscientific experiment, my blood pressure averaged 131.4/71.8 listening to the 440Hz version, 135.6/72.4 for the 432Hz re-tune. My heart rate was 64.2 at 440, 64.8 at 432. After the experiment, the readings were 138.7/65.3 and 65.7, respectively. As a scientist, I would label these results "not statistically significant". The variation among different readings under the same conditions was far greater than any difference between the higher and lower pitched listening versions (systolic blood pressure varied by 21 points during the 440Hz version, for instance, while the average difference of 440 to 432 was only 4.2). I can't draw any conclusions.

Still, it is the nature of the musical experience we are concerned with. From a purely subjective standpoint, I noted that when I listened to the Journey song at 432 immediately after the 440 version, I could tell a difference, even though it didn't show in my cardiovascular measurements; it actually felt a little more pleasant at the lower frequency. In general I prefer lower pitched instruments and those that generate fewer high pitched harmonics. "Don't Stop Believin" is in the key of E major, making the 432 version halfway between E and Eb major; I wonder how I would feel if the song was simply recorded in the key of Eb with the instruments in standard tuning.

For players of stringed instruments, the frequency to which they are tuned does make a difference. Assuming that the same type of strings are used on the violin, cello, or guitar, the tension on the strings would be slightly less at the lower frequency, making it easier for the fingers of the left hand to press down the notes. The slacker strings would also tend to produce more of the fundamental (the main note) and less of the higher overtones (the faint notes heard as the string vibrates at 1/2, 1/3, and 1/4 of its full length).

Theory is fine, but we want to know about how the musical experience feels to the fingers of the musician and the ears of the audience. I tuned one of my acoustic steel-string guitars down to A=342Hz (any of my guitars that I haven't played in a long time have already settled a quarter tone flat anyway) and compared it to my Fender acoustic steel-string that is tuned up to a full A=440Hz. The lower pitched guitar did feel a little easier to play, and sounded a little less bright. I asked my 13-year-old granddaughter, who is just getting into guitars, and she had the same impression.

I remain skeptical that music tuned to 432Hz can ease anxiety, improve cognition, or cure hypertension, but if twisting the tuners on my guitar a couple of times to the left might bring physical healing and peace on earth, why not?

Perfectly Imperfect: harmony and dissonance

A five-year-old girl, with a frilly dress and jet black hair pulled into two tight pony tails, sits down hesitantly on the piano bench while a dozen parents sit on folding chairs. This is Aisha's first piano recital, and her parents are more nervous than she is. Aisha tests out two notes, G and F together, one with each index finger. The notes seem to clash, but she repeats the tones five more times. Then her left finger moves one note to the left, to E; her listeners relax visibly as she plays this pair of notes six times. In a daring leap, her right finger moves up two notes, to B, while her left finger retreats one more key to the left, to D, five repeats, then as the adults in the chairs smile a little, each finger moves one space in opposite directions, to C and C, like stepping onto solid ground.

Aisha turns around and gives a little curtsey as she climbs down from the piano bench, and the adults applaud—her parents more enthusiastically than the other parents, who are waiting for their own darlings to show off what they have been practicing.

Piano recitals are a childhood rite of passage (or at least they were in the old days when every child took piano lessons, whether they wanted to or not). Aisha played "Chopsticks" as well as any other five-year-old. The magic of this starter piano piece comes from the relationships between the right hand (G, G, D, C) and the opposing left hand (F, E, D, C)—the magic of harmony, each combination evoking a different feeling and a different expectation in the ears of the listeners: Clashing dissonance with the first pair of notes, pleasant calming with the second pair, suggestive consonance with the third pair, and solid firmness with the final pair of notes, C to shining C.

The explanation for the sensations caused by each combination of notes starts with Pythagorus, ancient Greek mathematician, physicist, musician, and father of a spiritual movement based on the "music of the spheres".

The mathematical and musical observations of Pythagorus were taken as divine truth and became the basis of most Western music. Even when the math didn't quite work out (mathematical ratios work well when only one tonality is involved), compromises were made to preserve the relationships between notes as closely as possible (the "tempered scale".)

Musical instruments, like the piano, fretted string instruments, and most horns, were invented and built around the almost-perfect harmonies of scales drawn from twelve equally spaced steps. Harmony was in tune with an orderly universe.

But maybe Pythagorus was wrong.

Since music has been a part of every hominid culture, even before Homo sapiens, it was surely noticed that two notes sung or played simultaneously created a special feeling, and different combinations produced different feelings. Early humans must have known that when men chanted in a low voice and the women joined in their own higher register, the sound was richer, fuller. The apparent unity, but difference, must have sent shivers down the spines of these early musicians. Whether we count music's origins to a hundred centuries ago or back to pre-sapiens times, different voices produced different tones, blended in some sort of harmony. The forms that harmony has taken since have been analyzed, developed, complicated, and fought over ever since.

Pythagorus, when he wasn't defining the science of geometry, spent a lot of time mixing art and science (and religion) and thinking about why some notes sounded good together, others, not so much. Using a single-stringed instrument that he invented (patent still pending) called a monochord, Mr P. observed that when half the string vibrated by itself the note produced was an octave higher than when the whole string vibrated. The two notes sounded the same, but different, and this marked the beginning of music theory. If a third of the string was vibrated, the note was five steps (in our current scale system, at least) above the octave, and a quarter of the string produced two octaves above the whole string. Pythagorus considered both music and mathematics to be divine, so his conclusion, which has been passed down in some form for 2500 years, is that notes sound good together when the ratio between the lengths of a vibrating string is composed of small integers. Of course, he had no way of quantifying, or even conceiving of, the fact that the pitches were actually frequencies of pressure waves in the air, transmitted to the brain by rows of hairlike cells in the inner ear, each tuned to a different frequency. Pythagorus' mathematical/religious view established the concept of consonance (simple mathematical relationships of note intervals) and dissonance (more complex mathematical relationships) that is the basis for harmony found in Western music. Pythagorus' "music of the spheres" became part of his religious movement. Maybe we should reconsider.

First, there is the issue of dissonance. If two adjacent notes on a piano are played at the same time, lets say C and C#, the listener winces at the combination. There are several current theories about our reaction to dissonance. One theory notes that the cells of the middle ear are arranged in order of the frequencies that stimulate them, each note having a band of cells that responds to it. The cells in the middle of the band for middle C respond most strongly at 261.6Hz, but near the edges of the "critical band" for C the cells still fire, just less intensely. The cells of the critical band for C# that are centered at 277.2Hz lay right alongside, and the edges overlap. When C and C# are played together, cells of both frequencies fire, the brain struggles to process the information from the overlap area, and we perceive dissonance. An alternate theory of dissonance focuses on the interference of notes with nearly the same frequencies. Sound waves that are slightly out of sync create peaks and troughs that distort the pure tones, like the ocean surf in a crosswind which makes the waves too bumpy and gravelly for smooth surfing. The notes that clash are not close enough to create audible beats described in a previous chapter, but the gritty textures are harder for the brain to process.

Is that such a bad thing?

A team of researchers led by Dr. Peter Harrison tested the idea that people prefer sounds containing simple mathematical frequency ratios. They found that subjects were actually more drawn to harmonies that challenged numerical relationships, just slightly tilted from the Pythagorean ideal. They asked 4,000 subjects from the United States and Korea to listen to chords and rate how much they liked them. Then the participants used a slider to adjust certain notes until the chords sounded more pleasant. The results of 235,000 judgements revealed that people actually preferred harmonies that deviated from the mathematical. The effect was equally evident in both the American and Korean participants, suggesting a universal preference, regardless of musical culture.

It turns out that we like a little dissonance with our music. Our tonal system, and the instruments we have designed for it, tend to minimize dissonance, even though we like a little of it. How can we add a little imperfection to our traditional system of harmony?

French composer Claude Debussy was fond of using a major scale with the fifth raised a half step (a C major scale with G# instead of G). The dissonance created by shifting this one note created the dreamy nostalgic harmonic mist of Clair de Lune and other Impressionist compositions.

In the 20th century, classical composers and jazz improvisors found new ways to create dissonance. The chords of traditional Western tonal harmony use alternate steps of the scale—the root, 3rd, 5th, and sometimes the 7th. These notes happen to match the overtone series generated by strings and brass instruments. But why stop at the 7th? Interesting harmonic combinations occur if we go past the octave and keep skipping notes. Why not the 9th? The 11th? The 13th? After that, we are back at the root of the scale, but we have used all of the notes of the scale and found lots of interesting friction.

And if the chord extensions were not enough, each of the notes in the chord could be altered: the 9th might be flatted or sharped, the 11th could be raised, the 5th could be lowered or raised by a half step. Delicious dissonance in abundance.

Would Pythagorus have been fascinated or revolted? In early church music, the interval of a flatted 5th (also called a tritone, C to Gb) was forbidden, and it was termed the "devil's interval". And yet, in JS Bach's divine music we find him frequently using the diminished chord (root, flatted 3rd, flatted fifth, and double-flatted 7th), in which half of the intervals are those satanic tritones.

Blues and rock musicians have their own approach to dissonance. The chords are usually simple and mathematical, but the vocals and guitar leads bend the notes to conflict with the background harmonies. Electric guitars are often fitted with a "wang bar", a lever attached to the bridge for the sole purpose of pulling the strings out of tune. Pythagorus might be covering his ears.

Harmonies that don't depend on simple frequency ratios are not necessarily unnatural, but most of our Western instruments are designed to create more perfect intervals. Instruments from other cultures don't follow the same rules. In particular, percussion instruments that are struck to produce their tones often have overtones that ignore the acoustical rules of a plucked string.

Even the well-tuned bell that one might have heard in a church steeple in Mozart's home town would have an overtone series of 1:2, 1:2.367, 1:2.997, and 1:4, which Mozart's piano would have been unable to emulate.

Other harmonies and other scales with pleasing dissonances are found in other cultures around the world. Dr. Harrison's study cites the gamelan music of Indonesia, which includes a percussion instrument called the bonang. This bell-like instrument produces "inharmonic" overtones; its harmonic intervals lack simple mathematical relationships, but give rise to the scales and chords that give Indonesia's gamelan orchestras their ethereal beauty. Dr. Harrison suggests that music producers in all genres might explore instruments from other cultures (or invent their own) to broaden the palate of sounds that move us.

Although musical tastes can differ widely across the world, the inharmonic sounds of the bonang, bells, and gongs add human interest and life to the music. Our brains can find beauty in the slightly broken and out of tune. In Japanese aesthetics there is a concept termed wabi sabi, which stresses the beauty of irregularities in nature and life. Objects, as well as our human lives and the world around us, are beautiful precisely because they are imperfect.

Musicians should look for new instruments, new scales and harmonies, and new explanations for meaning in the universe.

Come Together: collective dynamics

Time is like a river, poets write. Moment to moment things change, hurry on, and are gone. The flow of music is much the same, and yet musicians who play together must keep together, as swimmers might grasp each other's hands while the current sweeps them away. The phenomenon of musicians coordinating their movements within milliseconds seems like magic.

I have always thought of the Beatles as creative song writers and ground-breaking teen idols, but not as finely tuned performers on par with orchestral violinists or tight-knit jazz ensembles. But Ringo Starr describes what it can be like to play music together: "It feels great; it's just a knowing. It's magic actually; it is pure magic. Everyone who is playing at that time knows where everybody's going. We all feel like one; where you go, everyone feels that's where we should go. I would know if Paul was going to do something, or if George was going to raise it up a bit, or John would double, or we'd bring it down… You've got to trust each other."

It is a sunny day with a pleasant breeze, and I am watching the San Diego Symphony perform John William's score to Star Wars at the Rady Shell, a magnificent outdoor venue on the edge of the bay. The movie video is playing on a huge screen near the edge of the stage, and the orchestra is perfectly in sync with the film as Luke and his friends battle Darth Vader. My attention is diverted from the screen to the perfectly coordinated motions of the violin section; every bow travels upward at exactly the same speed, and then descends in a flurry of notes together, moving and sounding as one. Their precision seems more amazing than the Millennium Falcon's jump to light speed on the screen. How can their movements be so closely linked?

Synchronization seems like magic, but it is common in our world. Fireflies flash their lights in the dusk, at first in random fashion, and then in tightly coordinated patterns. If two people are walking, especially if they know each other well, they will eventually "fall in step" with each other unconsciously. Physicists would call these systems, fireflies and feet, "coupled oscillators".

The first scientist to study the synching of movement was Christian Huygens, a Dutch physicist. He invented the pendulum clock, and later observed that when several of his clocks were in the same room, their pendulums would start to move exactly together; if he interrupted the movement, within a half hour they would again be swinging together.

Other scientists have explored naturally-occurring synchronization in the form of flocking birds. Using high-speed cameras and computer algorithms they study flocks of starlings, which rise when startled and fill the air with thousands of blackbirds flying in perfect synchrony as they make sharp dips and turns. The flocks are not following a leader, but moving as one. Although earlier observers imagined that there must be some sort of telekinesis at work, video analysis and computer simulation show that the phenomenon could be reproduced with a set of three simple rules: Maintain a constant distance from your neighbors, follow the average heading of those around you, and steer towards the average position of all of the others. And be quick and agile. Each individual is thus influenced by the nearest birds, but the number of "influencers" is limited to six or seven. Simple rules can create great complexity. The effects of a few rules on a large number of objects can create something like a miracle of nature.

The birds' limit of six or seven nearby guides seems surprisingly similar to the limits of human short-term memory, seven plus or minus one. A musician learning a new song by ear can only remember about seven notes, unless they can be grouped into smaller groups using some rule, such as "a descending scale of ten notes".

The grouping of six or seven birds in a large flock is also reminiscent of the size of each string section in an orchestra; perhaps those perfectly synchronized violin bows are partly a result of watching the two players on the right, on the left, and immediately in front. The conductor would like to pretend that he or she keeps the strings in time with their baton, but the notes move much faster than the conductor can direct. Watching the symphony play Star Wars, it seems like the string section is perfectly together, but it is hard to see the same precision between the musicians and the conductor.

Not all orchestras even need a conductor. One such conductor-less orchestra is the Orpheus Chamber Orchestra in New York City. With 70 albums to their credit, an international touring reputation, and collaborators like YoYo Ma and Isaac Stern, they are able to make world-class music by watching, feeling, and listening to each other.

In his scholarly book, “Beethoven’s Anvil: Music in Mind and Culture”, cognitive scientist and professional trumpet player William Benzon presents an elaborate argument that playing music together, which he calls “musicking”, influenced the development of the human brain and made us what we are. As early hominids left the safety of the African forest, Benzon suggests, the need for social adhesion became critical. Small groups of hunter-gatherers venturing onto the open savanna or the chilly steppes of receding ice ages could no longer afford to be independent or violently competitive like other primates. Primitive music, ritual dance, and chanting created bonds that kept bands together. Groups that shared a biological and cultural ability to make music together were better able to settle disputes and develop social relationships. Benzon pursues the idea that “group synchrony achieved through music and dance was an evolutionary precursor to language”, although the origins of both are an ongoing controversy. As evidence, the author describes primitive societies, where music even now plays a central role in the life of the tribe. As societies grew into civilizations, people had to work in groups, and music coordinated the workers as they chopped trees, hauled sails, and winnowed grain, together. The skills required were reinforced by sharing music; singing and dancing rituals developed in tandem with cooperative work. This concurrent development was not just cultural, according to Benzon. Our brains evolved for musical collaboration, and the neural functions of collaborators become coupled, forming biological systems with a musical life of their own. The essence of his theory is summarized by Freeman's Hypothesis: "By attending to one another through musicking, performers attune their nervous systems to each others'. This results in more harmonious interaction within the group." We share our minds through music.

This description of 'mind melding" starts to sound somewhat New Age and mystical, so it is time to look for scientific confirmation with a trip into some brain chemistry. There are the usual suspects: Endorphins make us feel good playing and listening to music, and dopamine motivates us. New insights involve oxytocin (sometimes nicknamed the "love hormone"), a peptide that suppresses previous learning by freeing the synaptic connections of neural networks. By loosening up the way we see things, we are able to form new bonds with the neural networks of others. Maybe not a mind meld, but closer communication.

There is also a similarity between dreaming and performing music with others. Among the dozens of chemical systems in the brain, two are very important: The aminergic system deals with learning and memory and is driven by serotonin and norepinephrine, while the cholinergic system uses acetylcholine to stimulate behavior and drive the motor pattern generators needed for walking and other rhythmic activities. While we are awake, both systems are in the game, keeping us thinking and moving. During normal sleep, both systems shut down to give us a break. But during dreaming, the cholinergic system is at waking levels (although movement is suppressed), while the aminergic system stays muted. Hence dreams can be full of mental activity, but we have a hard time remembering what happened after we wake. When making music together, the cholinergic system is in full gear (with motion no longer suppressed), but the aminergic chemicals are low, as they are during dreams. This may explain when we can be in the groove, interacting with other musicians, but afterwards our memories are fuzzy. If this sounds like some sort of trance state, it points back to the way that music and dance served ancient tribes in spiritual rituals that produced altered states among their members.

In Benzon's book he quotes Leonard Bernstein's description of a trance-like state that he sometimes experiences while leading an orchestra playing the music of a composer that he knows well: "You forget who you are or where you are and you *write the piece right there*. You just make it up as though you never heard it before. Because you *become* that composer. I always know when such a thing has happened because it takes me so long to come back. It takes four or five minutes to know what city I'm in, who the orchestra is, who are the people making all that noise behind me, *who am I?*"

When my son Tuffer was in his late teens, his musical tastes ran to rock and rap, the music of his peers. I talked him into coming to a jazz club to hear what jazz was all about. He had heard my jazz records around the house and wasn't particularly interested, but agreed to listen. At Jazz Alley, Joshua Redman was playing with his quartet, which included pianist Brad Mehldau, bassist

Christian McBride, and drummer Brian Blade. These musicians are now four of the premier players in jazz, but at the time they were “young lions”, early twenty-somethings on their first real tour as an ensemble, and they were obviously having a great time improvising together night after night. After we settled in, I noticed that my son was intently digging the music, caught up in every phrase. As we walked to the car after the show, he remarked at how the musicians reacted and responded to each other; “I heard the drummer play some little accent, and in the next beat the pianist was playing it back to him. I know they were improvising on the spot, but sometimes the sax and piano seemed to be playing exactly together, as if they had practiced it! How do they do that?” That is the difference, I told him, between hearing a recording and seeing jazz played live; when they are playing live, it is as close to telepathy as you can get. Tuffer came back the following night to listen to the band again, without me.

The musical coupling of nervous systems is not limited to performers.

A few weeks later Tuffer took me to a Bruce Springsteen concert at the Seattle Center Arena, where The Boss played for three and a half hours, non-stop. It was obvious that there was a neural coupling between the audience and the musicians on stage. The crowd was rocking, dancing, and yelling the lyrics to “Born in the USA!” in unison. Music was clearly a communal activity. The show was great, but by midnight I was exhausted and told my son it was time we went home—enough group bonding.

What we think of as our individual consciousness involves listening to our "inner voice", the language stream that constantly narrates what we are thinking and doing. Making music together by necessity silences the inner voice in order to hear the other musicians and add our own contribution. Silencing the inner voice makes room to join the neural activity of others, in what might be considered a form of telepathy, a state of flow. Music is a form of communal play that binds us to one another, biology, neurology, and society. And it gives us pleasure, together.

Contrapunctus: music as information, and god

The rich sound of a single violin in its lower range plays four notes from a G minor triad. As the melodic line starts to develop, a second violin enters with a figure similar to the first, but higher in the chord. A cello states the first figure in a lower register while the other voices are in motion, moving like tributaries of a stream which weave in and out, pausing and quickening, always in relationship to the other lines. Finally a viola, and we hear the unmistakeable perfection of "Contrapunctus 1" from Johann Sebastian Bach's "The Art of the Fugue". If there is divine music in the universe, this is it.

Bach's compositions are irresistible, with a gravitational force that draws in musicians and listeners of every genre. His compositions carry both infinite variation and seeming inevitability, reassuring us that the world is orderly, full of possibilities and, in the end, it all makes sense.

Millions of words have been spilled trying to explain why Bach's music has this effect. Musicologists analyze his use of harmony and counterpoint. Jazz musicians dig the way he can make an endless line of eighth notes flow effortlessly over a changing progression of harmonies. Philosophers see the transparent logic in his development of a single motif into a coherent argument. Musicians of all types, from bluegrass pickers to jazz improvisors and classical artists, use his preludes and inventions as practice vehicles to clear their minds and loosen their fingers. And church organists create spiritual spaces for worshippers, transcendent chords floating as light streams thru stained glass windows on Sunday morning. I propose that there is a common theme that explains the power of Bach, music, and existence: Information.

We think that we know what information is, but we often confuse information with communication, which involves the exchange of information. In the broadest sense, any system which is not homogenous contains information. When hot water is poured into one end of a bathtub of cold water, one end of the tub becomes warmer than the other and this temperature gradient is a form of information. While we might communicate to another person to "watch out, I just added hot water to that end of the tub", the information is there, whether we share our knowledge or not. Any given water molecule at the cold end of the tub might just happen to have the same energy and velocity as a molecule at the hot end, but statistically there are more fast molecules at one end and more slow molecules at the other, which is information.

In our theoretical bathtub the temperature will gradually even out as the fast molecules bump the cool ones and speed them up, or as we stir up the water to mix hot and cold. The loss of the water's organization is called entropy; individual bits of H2O are now moving randomly, the temperature of the water has become homogenous, and information has been lost.

In the 1940's, Claude Shannon was studying probability and prediction at Bell Labs when he wrote two scientific papers that established the basis for Information Theory. As often happens, this new way of looking at the world catalyzed fields as different as genetics, linguistics, and computers. And maybe even theology—we'll get to that later.

As information theory developed, a number of profound principles emerged. The first is that uncertainty is a necessary part of any system that has the potential for growth and meaning, and the greater the number of possibilities, the more powerful information becomes. When information is communicated, the entropy—what the recipient does not know—is reduced. Entropy is a matter of probability; if the number of possibilities is vast, entropy is high until information causes a collapse of the many possibilities into a single point of knowledge. The more possibilities there are, and the more unlikely the possibility is, the more information is conveyed.

In his insightful book, "Grammatical Man: Information, Entropy, Language, and Life", Jeremy Campbell explores the history and implications of information theory and poses a question that has puzzled philosophers for centuries: "Why are nature's products so improbable when they contain so much order?" Campbell traces the origin of the word "information" to medieval times, when information had a more active meaning: "Something which gives a certain form or character to matter, or to the mind: a force which shapes conduct, trains, instructs, inspires, or guides" Breaking apart the word, it becomes in-form-ation, that which forms-into, literally "that which creates". In theological terms, this becomes the Creator, and we start to glimpse what human intuition has tried to grasp with the concept of God. To put it into religious terms, a physicist's Trinity might might explain the universe as Information acting upon Matter and Energy.

Before we travel too far into the philosophical and theological woods, let's remember that we came here to talk about music. We are exploring the feeling that music can create, and we know that we are hearing and feeling something greater than the energy of vibrations in air, something with meaning; in the words of Victor Hugo, "Music expresses that which cannot be put into words and that which cannot remain silent."

If the power of music lies in the vast amount of information it contains, hinting at the very fabric of the universe, we need to look beyond the vibrating strings, the love songs, and the inked dots of a music manuscript. We will skip over the atomic levels of information contained in protons, neutrons, and electrons in their quantum levels (since we are talking physics, I was required to work in the word "quantum), and think about the vibration of air molecules. Of course, one might point out that the very fact that our atmosphere is made up of oxygen, nitrogen, and hydrogen is information by itself; as other planets offer possibilities of their own.

Air molecules in an empty concert hall might be evenly mixed, all at the same pressure. On stage a chair is moved, making a sound; the air vibrates, creating disordered differences in air pressure; information is generated, but it is simple and chaotic. A young woman with a violin takes her seat on the chair and tunes up, drawing her bow across the A string. Out of all of the possible movements of air in the room, these pulsations of air pressure occur 440 times a second and order is created from the random air currents. Superimposed on these 440 pulsations are the overtones, pressure waves with frequencies of 880 and 1318 every second. The violinist looks over the thicket of ink spots on the paper on her stand, marked "Sonata No. 1", and draws her bow over all 4 strings, sounding the notes G, D, Bb, and G together, which resonate in mathematical harmony, fundamentals and overtones superimposed in satisfying order to announce that out of all the possible combinations, this music will be in the tonality of G minor. The melody descends in a stepwise fashion for seven notes, further establishing a selected set of tonal relationships from all of the other aerial vibrations in the room and establishing a pattern of gradual movement from one note to the next. After 20 melody notes, a generous amount of information has been provided, turning the entropy of the air pressure in the hall into music. And then surprise: At the beginning of the second bar, the violin again plays across the top three strings together, but this time they are D, C, and F#. Something unexpected occurs in the harmony, intervals of a flatted 7th and a sharped 4th, the most dissonant combinations. Just as the music had settled into a predictable pattern, the least predictable of juxtapositions appear. The most unexpected occurrences create the most information, but also create new expectations. And there it is, eleven notes later in time, the same four notes that opened the piece. And so it goes, patterns repeated and broken for 21 bars and 528 notes. Thirty-five of the notes are sharped, breaking the predictability established by the key of G minor, while 32 are flatted and 39 are naturals that cancel the two flats of the key signature. Melody and harmony establish structures within the piece that weave in and out, creating order, expectation, and surprise, the building blocks of information. All in a minute and a half.

Bach's genius was that he could distill the range of musical possibilities with the generative power of rules, and yet make the rules transparent, leaving only the music, full of uncertainty and inevitability.

The violin performance touches only a small part of the information contained in the Sonata, however. Henry David Thoreau observed that "It takes two to speak the truth—one to speak and another to hear." Physicists speak of the "observer effect": theoretically, physical phenomena only become actual when someone sees them (like the cat in Schrodinger's famous thought experiment). Light itself acts like a wave if the experiment is measuring waves, but as a particle if searching for particles. Leaving aside the pointless if-a-tree-falls-in-a-forest question, music may express a vast amount of information when it is composed and performed, but an incomprehensible amount of information is created when the vibrations in the air reach the ears of a listener. We have traced many of the neural connections throughout this book, so we won't revisit the neurons and their networks again; simply note that the number of connected combinations in the brain exceeds the number of atoms in the universe. When Bach's music creates patterns in our neurons, it conjures one of an infinite number of other possible brain states. The inconceivable range of possibility seems to be very close to what we sense as divinity.

JS Bach's music is steeped in religion. He made his living working in German Lutheran churches as an organist, choirmeister, and composer, and many of his most famous works are explicitly Christian, like "Jesu, Joy Of Man's Desiring". His religious and secular compositions shared the same feeling, however, and the spiritual element of his music seems deeper, more profound, than the Lutheran theology that it supported or the church services where it was performed. I propose that what we hear in Bach's music has roots closer to God than to the church. Coming from generations of professional musicians, he was a master of his craft, but his music seems to reveal more profound truths, rather than simply accompanying religious services with music. The order of his compositions reveals the meaning-beyond-words inherent in logic, structure, uncertainty, and surprise, the essence of information.

The 20th century was marked with another musical genius whose deeply spiritual music came from a more unlikely background. John Coltrane was African-American, born in North Carolina in 1926. Twelve years later his father, aunt, and grandparents died in close succession and he and his mother moved to Philadelphia, He took band in high school and served in the military, but his musical development was shaped by playing tenor saxophone in big bands and blowing the blues in bar gigs. He played occasionally with Charlie Parker and Dizzy Gillespie, absorbing the complex harmonies of bebop jazz. He was mentored in the bands of innovators Miles Davis and Thelonious Monk, where zen-like simplicity was spiked with dissonance, before following his own search for deeper meaning and complexity with his own groups. What set Coltrane apart was his fanatical devotion to achieving the highest level possible. He practiced incessantly,"25 hours a day" according to a colleague, devouring patterns from violin exercise books and playing every possible variation of arpeggios in every key. He would sometimes fall asleep with his horn still in his mouth, and he took a noiseless saxophone on the airplane when he traveled to practice in the lavatory (times were different then!). His sound featured a tone so deep that it seemed to come from the earth itself, and he would squeeze every possible chord and scale into improvisations that would last for hours. His music, dubbed "sheets of sound", seemed to be an effort to mine every possibility that his horn could find, backed by pianist McCoy Tyner, who constructed thundering chords in intervals of 4ths to leave every harmonic possibility open, and drummer Elvin Jones, who created a polyrhythmic storm in which every rhythmic possibility was expressed simultaneously. Even though Coltrane's music was sometimes exhausting to listen to, one had the sense that it came from a spirituality that transcended explanation. While music critics focused on labels like "avant-garde" and "free jazz" to describe his quest for more complexity and boundary-breaking, Coltrane himself sought a unity based in simplicity and universality.

During the 1960's, when the country was struggling to find its own soul, Coltrane's music became more intense and explicitly spiritual, with titles like "Meditations", "Interstellar Space", and "Om", which opens with group chanting that evolves into a sea of simultaneous improvisation. The culmination of the saxophonist's quest is found in "A Love Supreme", a recording that has been widely regarded as one of the greatest record albums of all time. There is simplicity, a four-note theme that is repeated and elaborated into complex harmonic and rhythmic explorations by Coltrane's quartet.

Returning to the question behind all of these essays: "Why is music so meaningful?" This is a corollary to humanity's greatest questions: Why is there something, rather than nothing? Why are nature's products so improbable when they contain so much order? In a world where everything decays, how are even more complex new forms being created? How can something as infinitely complex as the human mind be created by the rules of the physical universe?

In 1977, Belgian physicist Ilya Prigonine received the Nobel Prize for his work on self-organizing systems. In his book, "Order Out Of Chaos", he explains that if there are enough varied possibilities in a system, new patterns and forms can come into being spontaneously, but only if energy is added to the system. Entropy degrades the information in a system, but energy and a set of rules (a form of information) can create forms of unlimited complexity.

Information Theory elaborates upon this principle: Below a certain "complexity barrier", entropy increases and systems become simpler. Once complexity rises above this critical level, however, more elaborate systems can develop rapidly.

The power of information can be seen in every aspect of the universe, from planetary motion to the DNA that creates life on Earth, from the structure of language to the possibilities of the human brain and mind. People (and we can't exclude our fellow creatures with nervous systems) have always felt wonder and awe at the order and complexity that can be generated by a world of uncertainty and chance.

Music's domain is the air that surrounds us; at every level, its vibrations express the power of information to create, soothe, inspire, and fill our minds with wonder. But information doesn't have to be "about" anything, which sets it apart from language. At the most basic level, music is pure information; Bach's "St. Mathew's Passion" is more profound than the story that it tells, Coltrane's "Psalm" from "A Love Supreme" is more than the result of obsessive saxophone practice, and the sounds of a bone flute and humans chanting around a campfire in the dawn of our species are more in touch with the origins and meaning of existence than our stories, our technologies, or our economies. From instruments fashioned of condor bones to the lyre-playing hero of ancient mythologies, from the baroque German High Church to cosmic jazz explorations, we can feel the inconceivable order and possibilities of the universe, whether we name it Information, Supreme Intelligence, or God.

Essayist Lewis Thomas said that if he could, he would send Bach's music out into the universe to explain us; "But that would just be bragging".

Epilogue:

Thank you for listening, thinking, and humming along with these musical ideas.

About The Author:

Dr. Lee R. Harris graduated from Washington State University College of Veterinary Medicine in 1974 and spent the next 45 years caring for dogs, cats, birds, and bunnies at his clinic near Seattle. He is also an enthusiastic amateur musician, playing a wide variety of instruments in jazz combos, R&B bands, church choirs, and community orchestras. Dr. Harris has published three books and an article in the Washington Post about animals and veterinary medicine, as well as a book of short biographies and quotes from lesser-known quotable personalities in history. He has explored other observations and experiences about music and the nervous system in his previous book of essays, titled "Searching for Orpheus".

In addition to his writing, Dr. Harris builds musical instruments and spends as much time as possible in the woods and mountains. His family and five granddaughters complete his life.

Acknowledgements:
I owe a debt to all of the people who have shared their music with me. Music is made of relationships, note-to-note and person-to-person. The friendship and forbearance of music teachers, bandmates, and audiences have enriched my life and taught me the many lessons that music has to offer. Credit is also owed to the infinite possibilities of the universe.

Cover by Reese Gordon, back cover picture by Tuffer Harris

Comments? lee@jazzydoc.com

Made in the USA
Monee, IL
04 November 2026

67381725R00098